I0093067

RANGER

HANDBOOK

*Not for the weak
or fainthearted*

SEPTEMBER 2025

This publication supersedes TC 3-21.76, dated 26 April 2017.

Headquarters, Department of the Army

This publication is available at the Army Publishing Directorate site (http://armypubs.army.mil) and the Central Army Registry site (https://atiam.train.army.mil/catalog/dashboard).

*TC 3-21.76

Training Circular
No. 3-21.76

Headquarters
Department of the Army
Washington, DC, 19 September 2025

RANGER HANDBOOK

Contents

Figures

Contents

Contents

Tables

This page intentionally left blank.

Preface

TC 3-21.76 provides a reference for students at the U.S. Army Ranger School. TC 3-21.76 is also a reference guide for small-unit tactics for combat arms units. Trainers and educators throughout the Army will also use this publication.

The principal audience for TC 3-21.76 is all members of the profession of arms. Commanders and staffs of Army headquarters serving as joint task force or multinational headquarters also refer to applicable joint or multinational doctrine concerning the range of military operations and joint or multinational forces.

Commanders, staffs, and subordinates ensure their decisions and actions comply with applicable United States, international, and, in some cases, host-nation laws and regulations. Commanders at all levels ensure Service members operate in accordance with the law of armed conflict and the rules of engagement. (See FM 6-27 for legal compliance.) Table 15-1 of TC 3-21.76 implements STANAG 3204, AAMedP-1.1. Paragraph 15-36 of TC 3-21.76 implements STANAG 2546, AJMedP-2.

TC 3-21.76 is not the proponent (the authority) of any Army terms. The use of a trade or brand name does not constitute an endorsement of any specific commercial product, commodity, service, or enterprise by the United States Army.

TC 3-21.76 applies to the Active Army, Army National Guard (of the United States), and United States Army Reserve unless otherwise stated.

The proponent of TC 3-21.76 is the United States Army Maneuver Center of Excellence. The preparing agency is the Department of Tactics, Training, and Doctrine, United States Army Maneuver Center of Excellence. Send comments and recommendations on DA Form 2028 (*Recommended Changes to Publications and Blank Forms*) to Commander, Airborne and Ranger Training Brigade, ATTN: ATSH-RB, 10850 Schneider Road, Fort Benning, GA 31905-4159; by email to usarmy.benning.mcoe.mbx.doctrine@army.mil; or submit an electronic DA Form 2028.

This page intentionally left blank.

Introduction

RANGER HISTORY

The history of the American Ranger is a long and colorful saga of courage, daring, and outstanding leadership. It is a story of warfighters whose skills in the art of fighting have seldom been surpassed. Only the highlights of their numerous exploits are told here.

Rangers mainly performed defensive missions until, during King Philip's War in 1675, Benjamin Church's Company of Independent Rangers from Plymouth Colony conducted raids on hostile American Indians. Upon the onset of the French and Indian War, provincial Captain Robert Rogers recruited nine companies of colonists to fight for the British. Ranger techniques and methods of operation were inherent to American frontiersmen. Rogers, who earned promotion to major through the process, became the first to capitalize on those and incorporate them into the fighting doctrine of a permanently organized force.

During the Revolutionary War, Colonel Daniel Morgan, who organized a unit known as Morgan's Riflemen, further developed the first Rangers' method of fighting. Francis Marion, the Swamp Fox, organized another famous Revolutionary War Ranger element known as Marion's Partisans. They numbered anywhere from a handful to several hundred, and they operated both with and independent of other elements of General George Washington's Army. Operating out of Carolinian swamps, they disrupted British communications and prevented the organization of Loyalists to support the British cause, substantially contributing to the American victory.

The U.S.-American Civil War again occasioned the creation of special units like Rangers. John Mosby, a master in the prompt, skillful use of cavalry, was one of the most outstanding Confederate Rangers. He believed resorting to aggressive action could compel enemies to guard a hundred points. He would then attack one of the weakest points, thereby ensuring numerical superiority.

With the entry of the United States into the Second World War, Rangers added to these pages of history. Major William Darby organized and activated the 1st Ranger Battalion on June 19, 1942, at Carrickfergus, Northern Ireland. The members all handpicked volunteers, 50 participated in the Dieppe Raid of the French coast with British and Canadian commandos. The 1st, 3rd, and 4th Ranger Battalions participated in the North African, Sicilian, and Italian campaigns. Darby's Ranger battalions spearheaded the Seventh Army landing at Gela and Licata during the Sicilian invasion and played a key role in the subsequent campaign, which ended in the capture of Messina. They infiltrated German lines and mounted an attack against Cisterna, where they annihilated a German parachute regiment during close-in, night, bayonet, and hand-to-hand fighting.

The participation of the 2nd and 5th Ranger Battalions in the D-Day landings of Operation Overlord at Omaha Beach inspired the words that would become the Rangers' official motto, "Rangers lead the way!" They spearheaded the drive inland to destroy the German gun emplacements trained on the beaches and enabled Allied forces to move away from their fire. In particular, the 5th Ranger Battalion scaled the cliffs of Pointe du Hoc.

The 6th Ranger Battalion, operating in the Pacific, conducted Ranger-type missions behind enemy lines of reconnaissance and long-range raids. It was the first U.S.-American group to return to the Philippines, destroying key coastal installations prior to the invasion. A reinforced company formed the rescue force who liberated U.S.-American and Allied prisoners of war from the Japanese prison camp at Cabanatuan.

Another Ranger-type unit of the Second World War was the 5307th Composite Unit (Provisional) under Major General Frank Merrill, after whom they came to be known as Merrill's Marauders. Volunteers from the 5th, 14th, and 33rd Infantry Regiments and from other Infantry regiments engaged in combat in the China-Burma-India Theater brought experience from jungle-trained and -tested units to learn and employ long-range penetration tactics for the jungle environment. The coordinated offensive campaign for Allied forces to gain control of north Burma employed them behind enemy lines in Japanese-controlled areas. In coordinated operations with the Chinese 22nd and 38th Divisions, they cleared the way for the construction of Ledo Road. They conducted missions in the jungles and mountains from Hukawng Valley to the Irrawaddy River. By disrupting Japanese supply lines and communications, they enabled Chinese southward movement. Their final victory was the capture of Myitkyina airfield, after which the unit disbanded in 1944. Remaining personnel merged into the 475th Infantry Regiment, which fought its last battle in 1945 at Loi-kang Ridge, China. The 475th Infantry Regiment is the progenitor of the 75th Ranger Regiment.

Soon after the Korean War started in June 1950, the 8th Army Ranger Company formed of volunteers from U.S.-American units in Japan. The company trained in Korea and distinguished itself in combat during the drive to the Yalu River, performing task force and spearhead operations. During the massive Chinese intervention of November 1950, this small, vastly outnumbered unit withstood five enemy assaults on its position.

In September 1950, a Department of the Army message called for volunteers to train as Airborne Rangers. Five thousand regular Army paratroopers from the 82nd Airborne Division volunteered. Nine hundred were chosen to form the first eight Airborne Ranger companies. Regular Army and National Guard Infantry division volunteers composed nine more companies. These 17 Airborne Ranger companies were activated and trained at Fort Benning, Georgia. Most received more training in the Colorado mountains.

In 1950 and 1951, some 700 men of the 1st, 2nd, 3rd, 4th, 5th, and 8th Airborne Ranger Companies fought to the front of every American Infantry division in Korea. Attacking by land, water, and air, these six Ranger companies raided, penetrated, and ambushed North Korean and Chinese forces. They were the first Rangers to make combat jumps. After the Chinese intervention, these Rangers were the first U.S.-Americans to recross the 38th parallel. The 2nd Airborne Ranger Company was the only African-American Ranger unit in the history of the U.S. Army. One in nine of the men of the six Ranger companies fighting in Korea died on the battlefields.

Other Airborne Ranger companies served with Infantry divisions in the United States, Germany, and Japan. These volunteers fought as members of line Infantry units in Korea. They volunteered for the Army, Airborne, and Rangers and for combat. The first to wear the coveted Ranger tab, these men are the original Airborne Rangers. One Ranger, Donn Porter, received the Medal of Honor posthumously. Fourteen Korean War Rangers rose to general officer. Dozens more became colonels, senior noncommissioned officers, and civilian leaders.

In 1951, Army Chief of Staff General J. Lawton Collins directed the extension of Ranger training to all Army combat units. He directed the Commandant of the Infantry School to establish a Ranger Department. This new department would develop and conduct the Ranger course of instruction. His goal was to raise the standard of training in all combat units. The program built on lessons learned from the Second World War and Korean War.

During the Vietnam Conflict, 14 Ranger companies served from the Mekong Delta to the demilitarized zone, conducting long-range reconnaissance and exploitation operations and providing combat information. Initially designated long-range reconnaissance patrol and later long-range patrol companies, these units received designations C through P (with the exception of any Juliet Company) Rangers, 75th Infantry.

After Vietnam, Army Chief of Staff General Creighton Abrams recognized the need for a highly trained and highly mobile reaction force. He activated the first battalion-sized Ranger units since the Second World War—the 1st and 2nd Battalions (Ranger), 75th Infantry. The 1st Battalion trained at Fort Benning and was activated February 8, 1974, at Fort Stewart, both in Georgia. The 2nd Battalion was activated on October 1, 1974. The 1st Battalion is now based at Hunter Army Airfield, Georgia. The 2nd Battalion is based at Fort Lewis, Washington.

In April 1980, members of C Company 1/75 supported Operation Eagle Claw, the attempt to rescue U.S.-American hostages in Iran. Although the mission failed, the tactics, techniques, and procedures developed by the company further enhanced Ranger capabilities.

General Abrams's farsighted decision and the combat effectiveness of the Ranger battalions were proven during Operation Urgent Fury, the invasion of Grenada, in 1983. The mission was to protect U.S. citizens and restore democracy. The Ranger battalions conducted a low-level airborne assault to seize the Point Salines airfield. They continued operations for several days, eliminating pockets of resistance and rescuing U.S.-American students. Due to this success, in 1984, the Department of the Army increased the strength of Ranger units to their highest level in 40 years. To do this, it activated another Ranger battalion and a Ranger Regimental Headquarters. After this 3rd Battalion (Ranger), 75th Infantry and Headquarters Company (Ranger), 75th Infantry were activated, Ranger units received over 2,000 Soldier assignments. On February 3, 1986, the 75th Infantry was renamed the 75th Ranger Regiment.

The first time the regimental headquarters and all three Ranger battalions deployed together was during Operation Just Cause in 1989. The 75th Ranger Regiment spearheaded the assault into Panama by conducting airborne assaults on the Torrijos-Tocumen Airport and Rio Hato Airfield. Its mission was to facilitate the restoration of democracy in Panama and to protect the lives of U.S. citizens. Between December 20, 1989 and January 7, 1990, the regiment performed many follow-on missions in Panama.

Elements of the 75th Ranger Regiment deployed to Saudi Arabia to support Operation Desert Storm in 1991; to Somalia to support Operation Restore Hope in 1993; to Haiti to support Operation Uphold Democracy in 1994; and to Kosovo to support Operation Joint Guardian in 2000 and 2001.

Since September 11, 2001, the 75th Ranger Regiment has led the way in establishing democracy. In 2001, elements of the regiment deployed to Afghanistan to support Operation Enduring Freedom. In 2003, elements deployed to support Operation Iraqi Freedom. The regiment spearheaded the campaign against senior-level members of Al Qaeda and the Taliban in both theaters during the deployment of U.S. Forces to Afghanistan and Iraq. Additionally, Ranger leaders throughout the entire force participated in more than 15 years of combat and advisor (or advise and assist) operations, serving as points of continuity among a transforming Army.

Rangers continue to contribute significantly to the overall success of operations. The 75th Ranger Regiment stands ready to execute its special operations mission in support of U.S. policies and objectives. Rangers throughout the force lead their formations, set the example for fellow Soldiers, and remain ready to defend the United States against all enemies. Rangers lead the way!

RANGER CREED

Army Chief of Staff General Creighton Abrams ordered the formation of the Ranger battalions in 1974, and he directed that they would set the standards for the Army. Command Sergeant Major Neal Gentry wrote the Ranger Creed, which maintains a code of ethics and philosophy for Rangers to adopt and adapt and which embraces the spirit, discipline, and duty that are the hallmarks of all Rangers, whether in times of peace or of war.

RANGER CREED

- **R**ecognizing that I volunteered as a Ranger, fully knowing the hazards of my chosen profession, I will always endeavor to uphold the prestige, honor, and high esprit de corps of my Ranger Regiment.

- **A**cknowledging the fact that a Ranger is a more elite soldier, who arrives at the cutting edge of battle by land, sea, or air, I accept the fact that as a Ranger, my country expects me to move further, faster, and fight harder than any other soldier.

- **N**ever shall I fail my comrades. I will always keep myself mentally alert, physically strong, and morally straight, and I will shoulder more than my share of the task, whatever it may be, one hundred percent and then some.

- **G**allantly will I show the world that I am a specially selected and well trained soldier. My courtesy to superior officers, neatness of dress, and care of equipment shall set the example for others to follow.

- **E**nergetically will I meet the enemies of my country. I shall defeat them on the field of battle for I am better trained and will fight with all my might. Surrender is not a Ranger word. I will never leave a fallen comrade to fall into the hands of the enemy and under no circumstances will I ever embarrass my country.

- **R**eadily will I display the intestinal fortitude required to fight on to the Ranger objective and complete the mission, though I be the lone survivor.

ROGERS'S RANGERS

In the 1750s, Major Robert Rogers drafted 28 rules of discipline for his ranging school training plan. They served as a manual on guerrilla warfare for Rogers's Ranger company, comprising 600 members selected by Rogers. A little over 200 years later, Army Ranger doctrine adopted and incorporated a more popular rendition—based in the 1930s on Rogers's original tactics—into 19 rules, which stand to this day as part of Rangers legend.

STANDING ORDERS

- Don't forget nothing.

- Have your musket clean as a whistle, hatchet scoured, sixty rounds powder and ball, and be ready to march at a minute's warning.

- When you're on the march, act the way you would if you was sneaking up on a deer. See the enemy first.

- Tell the truth about what you see and what you do. There is an army depending on us for correct information. You can lie all you please when you tell other folks about the Rangers, but don't never lie to a Ranger or officer.

- Don't never take a chance you don't have to.

- When we're on the march we march single file, far enough apart so one shot can't go through two men.

- If we strike swamps, or soft ground, we spread out abreast, so it's hard to track us.

- When we march, we keep moving till dark, so as to give the enemy the least possible chance at us.

- When we camp, half the party stays awake while the other half sleeps.

- If we take prisoners, we keep 'em separate till we have had time to examine them, so they can't cook up a story between 'em.

- Don't ever march home the same way. Take a different route so you won't be ambushed.

- No matter whether we travel in big parties or little ones, each party has to keep a scout 20 yards ahead, 20 yards on each flank, and 20 yards in the rear so the main body can't be surprised and wiped out.

- Every night you'll be told where to meet if surrounded by a superior force.

- Don't sit down to eat without posting sentries.

- Don't sleep beyond dawn. Dawn's when the French and Indians attack.

- Don't cross a river by a regular ford.

- If somebody's trailing you, make a circle, come back onto your own tracks, and ambush the folks that aim to ambush you.

- Don't stand up when the enemy's coming against you. Kneel down, lie down, hide behind a tree.

- Let the enemy come till he's almost close enough to touch. Then let him have it and jump out and finish him up with your hatchet.

This page intentionally left blank.

Chapter 1

Leadership

An Army leader is anyone who, by virtue of assumed role or assigned responsibility, inspires and influences people to accomplish organizational goals. Army leaders motivate people inside and outside the chain of command to pursue actions, focus thinking, and shape decisions for the greater good of the organization. Leadership is the process of influencing people by providing purpose, direction, and motivation to accomplish the mission and improve the organization.

PRINCIPLES

1-1. Leadership, the most essential element of combat power, gives purpose, direction, and motivation in combat. The leader balances and maximizes maneuver, firepower, and protection against the enemy. This chapter discusses how a leader does this by exploring the principles of leadership; the duties, responsibilities, and actions of an effective leader; and the leader's assumption of command.

1-2. The leadership requirements model establishes what leaders need to be, know, and do. (See table 1-1.) A core set of requirements informs leaders about expectations. Leaders' adherence to these requirements sets them on the process to influencing people by providing purpose, direction, and motivation to accomplish the mission and improve the organization.

Table 1-1. Leadership requirements model

ATTRIBUTES		
Character Army values Empathy Warrior ethos Service ethos Discipline	**Presence** Military and professional bearing Fitness Confidence Resilience	**Intellect** Mental agility Sound judgment Innovation Interpersonal tact Expertise
Leads Leads others. Builds trust. Extends influence beyond the chain of command. Leads by example. Communicates.	**Develops** Creates a positive environment and fosters esprit de corps. Prepares self. Develops others. Stewards the profession.	**Achieves** Gets results.

Table 1-1. Leadership requirements model (continued)

COMPETENCIES			
Oath to Constitution. Subordinate to law and civilian authority.	Combat power: unifier and multiplier.	Influence: commitment, compliance, and resistance.	Positive and harmful forms of leadership.
LEVELS OF LEADERSHIP	Direct – refine ability to apply competencies at a proficient level. Organizational – apply competencies to increasingly complex situations. Situational – adjust actions to complex and uncertain environments.		
SPECIAL CONDITIONS OF LEADERSHIP	Formal – designated by rank or position. (Command is an example.) Informal – initiative taken and appropriate special expertise applied. Collective – synergy achieved with multiple leaders aligned by purpose. Situational – actions adjusted to complex and uncertain environments.		
OUTCOMES			
Secured United States interests. Mission success. Sound decisions.	Expertly led organizations. Stewardship of resources. Stronger families.		Fit units. Healthy climates. Engaged Soldiers and Civilians.

1-3. To complete all assigned tasks, all the Rangers in a patrol do their jobs. Each accomplishes specific duties and responsibilities and is part of the team. This includes the platoon leader (PL), platoon sergeant (PSG), squad leader (SL), weapons squad leader (WSL), team leader (TL), medic, radio operator, and forward observer (FO).

PLATOON LEADER

1-4. The PL is responsible for what the patrol does or fails to do. This includes tactical employment, training, administration, personnel management, and logistics. The PL manages these responsibilities by planning, making timely decisions, issuing orders, assigning tasks, and supervising patrol activities.

1-5. The PL knows the Rangers in the platoon and how to employ the patrol's weapons. The PL is responsible for positioning and employing all assigned or attached crew-served weapons and for the employment of supporting weapons. The PL also—

- Establishes the time schedule using reverse planning and incorporates time for execution, movement to the objective, and the planning and preparation phase of the operation.
- Takes the initiative to accomplish the mission in the absence of orders. Keeps higher headquarters (HQ) informed through periodic situation reports.
- Plans with the help of the PSG, SLs, and other key personnel such as TLs, FOs, and attachment leaders.
- Stays abreast of the situation by coordinating with adjacent patrols and higher HQ, supervises and issues fragmentary orders (FRAGORDs), and accomplishes the mission.
- Requests any additional requisite support for performing the patrol's mission from higher HQ.
- Directs and assists the PSG in planning and coordinating the patrol's sustainment effort and casualty evacuation (CASEVAC) plan.
- Receives on-hand status reports from the PSG and SLs during planning.

- Reviews patrol requirements based on the tactical plan.
- Ensures the continuous maintaining of all-around security
- Supervises and spot-checks all assigned tasks and corrects unsatisfactory actions.
- Maintains a position to influence the most critical task for mission accomplishment, usually with the main effort, to ensure the platoon achieves its decisive point during execution
- Commands through the SLs within the intent of the commanders two levels higher
- Conducts rehearsals.

PLATOON SERGEANT

1-6. The PSG is the senior noncommissioned officer in the patrol and second in succession of command. The PSG helps and advises the PL, leads the patrol in the leader's absence, supervises the patrol's administration, logistics, and maintenance, and prepares and issues paragraph 4 of the operation order (OPORD). The PSG also—

- Organizes and controls the patrol command post in accordance with the unit's standard operating procedure (SOP), the PL's guidance, and the METT-TC (I) variables: mission, enemy, terrain and weather, troops and support available, time available, civil considerations, and informational considerations
- Maintains patrol status of personnel, weapons, and equipment. Consolidates and forwards the patrol's casualty reports through DA Form 1156 (*Casualty Feeder Card*) and receives and orients replacements.
- Monitors the morale, discipline, and health of patrol members
- Supervises and directs the patrol's medic and aid and litter teams in moving casualties to the rear.
- Ensures the CASEVAC plan is complete and executed properly.
- Supervises task-organized elements of patrol. These include the following:
    - Quartering parties.
    - Security forces during withdrawals
    - Support elements during raids or attacks.
    - Security patrols during night attacks.
- Receives the SL's requests for rations, water, and ammunition. Works with the company first sergeant or executive officer to request resupply. Directs the routing of supplies and mail.
- Coordinates and supervises the patrol's resupply operations.
- Ensures supplies are distributed in accordance with the patrol leader's guidance and direction.
- Ensures ammunition, supplies, and loads are properly and evenly distributed (a critical task during consolidation and reorganization).
- Ensures the patrol adheres to the PL's time schedule.
- Assists the PL in supervising and spot-checking assigned tasks and in correcting unsatisfactory actions.

1-7. During movement and halts, the PSG takes necessary actions to facilitate movement, supervises rear security during movement; establishes, supervises, and maintains security during halts; performs additional tasks as required by the PL; and assists in every way possible. The PSG also focuses on security and control of the patrol

1-8. At <u>danger areas</u>, the PSG directs the positioning of near side security (usually conducted by the trail squad or team) and maintains accountability of personnel. During <u>actions on the objective</u>, the PSG—

- Assists with objective rally point occupation.
- Supervises, establishes, and maintains security at the objective rally point.
- Supervises the final preparation of Soldiers, weapons, and equipment in the objective rally point in accordance with the PL's guidance.
- Assists the patrol leader in control and security.
- Supervises the consolidation and reorganization of ammunition and equipment.
- Establishes marks, supervises the planned casualty collection point (CCP), and ensures personnel status including the accurate reporting of wounded in action or killed in action (KIA) to higher HQ.
- Performs additional tasks assigned by the PL and reports status.

1-9. During <u>actions in the patrol base</u> (PB), the PSG assists in PB occupation and in establishing and adjusting the perimeter. The PSG enforces security in the PB, keeps movement and noise to a minimum, supervises and enforces camouflage, assigns sectors of fire, and ensures designated personnel remain alert and equipment maintenance remains at a high state of readiness.

1-10. The PSG requisitions supplies, water, and ammunition and supervises their distribution. The PSG supervises the priority of work, ensures its accomplishment, and performs additional tasks assigned by the PL. When creating a <u>security plan</u>, the PSG ensures—

- Crew-served weapons have interlocking sectors of fire.
- Claymore mines' emplacements cover dead space.
- Range cards and sector sketches are complete. The security plan also includes the following:
  - Alert plan.
  - Evacuation plan.
  - Withdrawal plan.
  - Alternate PB.
  - Maintenance plan.
  - Hygiene plan.
  - Messing plan.
  - Water plan.
  - Rest plan.

SQUAD LEADER

1-11. The SL is responsible for what the squad does or fails to do. The SL is a tactical leader who leads by example. In addition to completing casualty feeder cards and reviewing the casualty reports completed by squad members, the SL—

- Directs the maintenance of the squad's weapons and equipment.
- Inspects the condition of Rangers' weapons, clothing, and equipment.
- Keeps the PL and PSG informed on the status of the squad.
- Submits the liquids, ammunition, casualties, and equipment (LACE) report to the PSG.

1-12. During <u>actions throughout the mission</u>, the SL obtains status reports from the TLs, submits those to the PL and PSG, makes recommendations to the PL or PSG about observed problems, delegates priority tasks to the TLs, and supervises their accomplishment in compliance.

1-13. The SL uses initiative in the absence of orders, follows the PL's plan, and makes recommendations. During <u>movement and halts</u>, the SL—

- Ensures members rotate heavy equipment and share difficult duties.
- Notifies the PL of the status of the squad.
- Maintains proper movement techniques while monitoring route, pace, and azimuth.
- Ensures the squad maintains security throughout the movement and at halts.
- Prevents breaks in contact.
- Ensures subordinate leaders are disseminating information, assigning sectors of fire, and checking personnel.

1-14. During <u>actions on the objective</u>, the SL ensures special equipment is ready for actions at the objective and maintains positive control of the squad during the execution of the mission. The SL also positions key weapon systems during and after assault on the objective, obtains status reports from the TLs, ensures the redistribution of ammunition, and reports the status to the PL. When performing <u>actions in the PB</u>, the SL—

- Ensures the PB is occupied in accordance with the plan.
- Ensures interlocking fires coverage over the entire sector of the PB and makes any necessary final adjustments.
- Sends out listening posts or observation posts (OPs) in front of the assigned sector depending on the METT-TC (I) variables.
- Ensures the accomplishment of the priorities of work and reports accomplished priorities to the PL and PSG.
- Adheres to the time schedule.
- Ensures personnel know the alert and evacuation plans and the locations of key leaders, OPs, and the alternate PB.

WEAPONS SQUAD LEADER

1-15. The WSL is responsible for all the weapons squad does or fails to do. The WSL's duties are the same as those of the SL. The WSL also controls the machine guns in support of the patrol's mission and advises the PL on employment of the squad. The WSL also—

- Supervises machine gun teams to ensure they follow the priorities of work.
- Inspects machine gun teams for correct range cards, fighting positions, and their understanding of the fire plan.
- Supervises the maintenance of machine guns, ensuring the correct performance of maintenance, the correction and reporting of deficiencies, and maintenance performance is not in violation of the security plan.
- Assists the PL in planning.
- Positions at halts and danger areas any machine guns not attached to squads in accordance with the patrol's SOP.
- Rotates loads. (Machine gunners normally get tired first.)

- Submits LACE report to the PSG.
- Designates sectors of fire, the principal direction of fire, and secondary sectors of fire for all guns.
- Gives fire commands to achieve maximum effectiveness of firepower. Commands include those to—
    - Shift fires.
    - Correct windage or elevation to increase accuracy.
    - Alternate firing guns.
    - Control rates of fire and fire distribution.
- Knows locations of assault and security elements and prevents fratricide.
- Reports to the PL.

TEAM LEADER

1-16. The TL controls the movement of the fire team and the rate and placement of fire. To do this, the TL leads from the front and uses the proper commands and signals. The TL maintains accountability of Rangers, weapons, and equipment and ensures Rangers maintain unit standards in all areas and are knowledgeable of their tasks and the operation.

1-17. The TL leads by example and has specific duties and responsibilities during <u>mission planning and execution</u>. The TL takes specific action when orders are issued; specifically, for—
- A warning order (WARNORD) by—
    - Assisting in control of the squad.
    - Monitoring the squad during the issuance of the order.
- An OPORD's preparation by—
    - Posting changes to the schedule.
    - Posting and updating team duties on the WARNORD board.
    - Submitting ammunition and supply requests.
    - Picking up the ammunition and supplies.
    - Distributing the ammunition and special equipment.
    - Performing all the given tasks in the SL's special instruction paragraph.
- An OPORD's issuance and rehearsal by—
    - Monitoring the squad during the issuance of the order.
    - Assisting the SL during rehearsals.
    - Taking necessary actions to facilitate movement.
    - Enforcing rear security.
    - Establishing, supervising, and maintaining security.
    - Performing other tasks as the SL requires and helping in every way, particularly in control and security.

1-18. During <u>action in the objective rally point</u>, the TL assists in the occupation of the point and helps supervise, establish, and maintain security. The TL also—
- Supervises the final preparation of Rangers, weapons, and equipment in the objective rally point in accordance with the SL's guidance.
- Assists in control of personnel departing and entering the objective rally point.
- Reorganizes the perimeter after the reconnaissance party departs.
- Maintains communication with higher HQ.

- Upon the return of the reconnaissance party, helps reorganize personnel and redistribute ammunition and equipment and ensures the maintaining of accountability of all personnel and equipment.
- Disseminates priority intelligence requirements (PIRs) to the team.
- Performs additional tasks assigned by the SL.

1-19. During actions on the objective, the TL ensures special equipment is ready and maintains positive control of the fire team. The TL also controls the movement and maneuver of individual fire team members and assists in positioning key weapon systems during and after the assault on the objective. During actions in the PB, the TL—

- Inspects the perimeter to ensure the team has interlocking sectors of fire.
- Prepares the team sector sketch.
- Enforces the priority of work and ensures the team accomplishes it properly.
- Upon the return of the reconnaissance party, helps reorganize personnel and redistribute ammunition and equipment and ensures the maintaining of accountability of all personnel and equipment.
- Disseminates the PIRs to the team.
- Performs additional tasks assigned by the SL and assists in every way possible.

MEDIC

1-20. The medic assists the PSG in directing the aid and litter teams and monitors the health and hygiene of the platoon. In addition to treating casualties, conducting triage, and assisting in CASEVACs under the control of the PSG, the medic—

- Aids the PL or PSG in field hygiene matters, personally checking the health and physical conditions of platoon members.
- Requests Class VIII (medical materiel) supplies through the PSG.
- Provides technical expertise to combat lifesavers (CLSs) and supervises them.
- Ensures casualty feeder cards are correct and attached to each evacuated casualty.
- Carries out other tasks assigned by the PL or PSG.

RADIO OPERATOR

1-21. The radio operator is responsible for establishing and maintaining communications with higher HQ and within the patrol. During planning, the radio operator—

- Enters the net at the specified time.
- Ensures all frequencies, communications security fills, and net identifications are preset in squad and platoon radios.
- Informs the SL and PL of changes to call signs, frequencies, challenges and passwords, and number combinations based on the appropriate time in the automated net control device.
- Ensures proper functioning of all radios and troubleshoots and reports deficiencies to higher HQ.
- Weatherproofs all communications equipment.

1-22. In addition to serving as an en route recorder during all phases of the mission, the radio operator—

- Tracks time after the initiation of the assault.
- Records all enemy contact and reports it to higher HQ in the size, activity, location, unit, time, and equipment (SALUTE) format.
- Reports all operation schedules to higher HQ.

FORWARD OBSERVER

1-23. The FO works for the PL, serving as the eyes and ears of field artillery and mortars. The FO is mainly responsible for locating targets and calling for and adjusting indirect fire support. The FO knows the terrain wherein the platoon is operating, tactical situation, mission, concept of operations, and unit's scheme of maneuver and priority of fires.

1-24. During planning and while preparing and using situation maps, overlays, and terrain sketches, the FO selects targets to support the platoon's mission based on the company OPORD, PL's guidance, and analysis of the METT-TC (I) variables. During execution, the TL—

- Informs the fire support team HQ of platoon activities and the fire support situation.
- Selects new targets to support the platoon's mission based on the company OPORD, the PL's guidance, and analysis of the METT-TC (I) variables.
- Calls for and adjusts fire support.
- Operates as a team with the radio operator.
- Selects OPs.
- Maintains communications as prescribed by the fire support officer.
- Maintains the eight-digit coordinate of the current location at all times.

ASSUMPTION OF COMMAND

1-25. Platoon and squad members take command of their elements upon occasion in an emergency, so every Ranger prepares to do so. During an assumption of command and when the situation permits, the Ranger assuming command accomplishes the tasks in table 1-2, though not necessarily in that order and based on the METT-TC (I) variables.

Table 1-2. Tasks for the assumption of command

INFORMS	The unit's subordinate leaders of the command and notifies higher headquarters.
CHECKS	Security.
CHECKS	Crew-served weapons.
PINPOINTS	Location.
COORDINATES AND CHECKS	Equipment.
CHECKS	Personnel status.
ISSUES	Requisite fragmentary orders.
REORGANIZES	On a needs basis, maintaining unit integrity when possible.
MAINTAINS	Noise and light disciplines.
CONTINUES	Patrol base activities, especially security, when assuming command in a patrol base.
RECONNOITERS	At the least, for map reconnaissance.
FINALIZES	Plan.

This page intentionally left blank.

Chapter 2

Operations

This chapter provides techniques and procedures that Infantry platoons and squads use throughout the planning and execution phases of tactical operations. Specifically, it discusses the necessary troop leading procedures (TLP), combat information, combat orders, and planning techniques and tools to prepare a platoon for combat operations. These topics are time sensitive and apply to all combat operations. When time permits, leaders plan and prepare in depth. When less time is available, leaders rely on previously rehearsed actions, battle drills, and SOPs.

TROOP LEADING PROCEDURES

2-1. A leader completes specific steps in the TLP (see table 2-1) to prepare the unit to accomplish a tactical mission.

Table 2-1. Steps in the troop leading procedures

1. Receive the mission.	5. Conduct reconnaissance.
2. Issue a warning order.	6. Complete the plan.
3. Make a tentative plan.	7. Issue the operation order.
4. Initiate movement.	8. Supervise and refine.

2-2. The TLP start when the leader is alerted for a mission or receives a new mission or a change to an existing mission. Steps 3 through 8 are done in any order or at the same time.

STEP 1–RECEIVE THE MISSION

2-3. The leader receives the mission in an OPORD or a FRAGORD. The 1/3 :2/3 rule only applies to the planning and preparation of an operation. Parallel planning occurs as the leader uses one-third of the available planning and preparation time while subordinates use the other two-thirds. Emphasis remains on conducting a hasty analysis with the primary focus on planning and preparation.

STEP 2–ISSUE A WARNING ORDER

2-4. The leader provides initial instructions in a WARNORD that contains enough information to begin preparations as soon as possible. The WARNORD mirrors the five-paragraph OPORD format. A WARNORD includes this essential information:

- Type of operation.
- General location of the operation.
- Initial operational timeline with emphasis on critical times.
- Reconnaissance to initiate.
- Movement to initiate.

- Planning and preparation instructions including a planning timeline.
- Information requirements.
- Commander's critical information requirement(s).

STEP 3–MAKE A TENTATIVE PLAN

2-5. During the planning process, leaders analyze the mission variables of METT-TC (I).

    a. Conduct a detailed mission analysis.

        (1) <u>Mission, intent, and concept</u> of higher commanders' concepts and intents two levels up. Find this information in the OPORD, paragraph 1b for two levels up and paras 2 and 3 for one higher up.

        (2) <u>Unit tasks</u>. These tasks are either clearly stated in the order (specified tasks) or become apparent during analysis of the OPORD (implied tasks). (See table 2-2.)

Table 2-2. Examples of specified and implied tasks

SPECIFIED TASKS	IMPLIED TASKS
Retain Hill 545. Provide one squad to the 81-millimeter platoon to carry ammunition. Establish an observation post, vicinity GA124325 not later than 301500 November 16.	Provide security during movement. Conduct resupply operations. Coordinate with adjacent units.

        (3) <u>Unit constraints</u>. The leader identifies any constraints placed on the unit. Constraints either prohibit or require an action. Leaders identify all constraints an OPORD places on their units' ability to execute their missions. The two types of constraints are prescriptive (that is, required, mandating action) and prohibitive (that is, forbidden, limiting action).

        (4) <u>Mission-essential task(s)</u>. After reviewing all the factors covered in previous paragraphs, the leader identifies the mission-essential task(s). Failure to accomplish such a task equals failure to accomplish the mission. The mission-essential task appears in the maneuver paragraph.

        (5) <u>Restated mission</u>. This clearly and concisely states the mission (the purpose to be achieved) and the requisite mission-essential task(s) for achieving it. It identifies <u>who</u>, <u>what</u> (offensive or defensive task and either action by friendly forces or effect on enemy forces), <u>when</u> (the critical time), <u>where</u> (usually grid coordinate), and <u>why</u> (the purpose the unit is to achieve nested both vertically and horizontally). (See table 2-3.)

Table 2-3. Examples of restated missions

(Who?) 1st Platoon attacks. *(What?)* To seize. *(Where?)* Hill 482, vicinity NB457371, Objective Blue. *(When?)* NLT 090500Z December 16. *(Why?)* IOT enable the company's main effort to destroy enemy command bunker.
(Who?) 1st Platoon, C Company defends. *(What?)* To destroy from. *(Where?)* AB163456 to AB163486 to AB123486 to AB123456. *(When?)* NLT 181530Z October 16. *(Why?)* IOT prevent enemy forces from enveloping B Company, 3-187th Infantry from the south.
Legend: IOT–in order to; NLT–not later than

b. Analyze the situation and develop a course of action (COA).

   (1) Feasible. A COA allows mission accomplishment within the given time, space, and resource limitations.

   (2) Acceptable. The military advantage gained by executing the COA justifies the cost in resources, especially casualties. This assessment is largely subjective.

   (3) Suitable. A COA is executable within the commander's intent and planning guidance.

   (4) Distinguishable A COA differs significantly enough from other possible COAs.

   (5) Complete. A COA incorporates how the decisive operation leads to mission accomplishment, how shaping operations create and preserve conditions for success of the decisive operation, and how sustaining operations enable shaping and decisive operations.

c. With the restated mission from Step 1 to provide focus, the leader continues the estimation process using the remaining METT-TC (I) variables:

   (1) Enemy. What is known about the enemy? (See table 2-4.)

Table 2-4. Knowledge of enemy

COMPOSITION	An analysis of the forces and weapons the enemy brings to bear including what weapon systems they have available and what additional weapons and units are supporting them.
DISPOSITION	How the enemy is arrayed on the terrain (such as in defensive positions) or in an assembly area, or how they are moving in march formation.
STRENGTH	The percentage of the strength and number of the passengers.
RECENT ACTIVITIES	Identification of recent and significant enemy activities that indicate future intentions.
REINFORCEMENT CAPABILITIES	Determination of positions for reserves and an estimated time to counterattack or reinforce.
POSSIBLE COURSES OF ACTION	Determination of the enemy's most likely and dangerous courses of action, the analysis of which decreases the likelihood of surprise to the friendly unit during execution.

   (2) Terrain and weather. How will terrain and weather affect the operation? Analyze terrain using observation and fields of fire, avenues of approach (AAs), key terrain, obstacles, and cover and concealment. (See table 2-5.)

Table 2-5. Offensive considerations

FRIENDLY	ENEMY
How can these avenues support my movement? What are the advantages and disadvantages of each? (Consider enemy, speed, cover, and concealment.) What are the likely enemy counterattack routes?	How can the enemy use these approaches? Which avenue is most dangerous? Least? (Prioritize each approach.) Which avenues support a counterattack?

(a) <u>Observation and fields of fire.</u> Determine locations that provide the best observation and fields of fire along the approaches, near the objective, or on key terrain. The main concern of the analysis of fields of fire is with the ability to cover the terrain with direct fire.

(b) <u>AAs.</u> Develop these next and identify those one level down. Also consider aerial and subterranean avenues. Use table 2-5 for offensive considerations to AAs.

(c) <u>Key terrain.</u> This is any location or area whose seizure, retention, or control affords a marked advantage to either combatant. Using the map and information already gathered, look for key terrain that dominates AAs or the objective area. Next, look for decisive terrain that has an extraordinary impact on the mission when held or controlled.

(d) <u>Obstacles.</u> Identify existing and reinforcing obstacles and hindering terrain that affect mobility.

(e) <u>Cover and concealment.</u> This analysis is often inseparable from that for the fields of fire and observation. Weapon positions are ideally both effective and survivable. Infantry units improve poor cover and concealment by digging in and camouflaging their positions. When moving, units use the terrain to provide cover and concealment.

(3) Troops and support available. Analyze troop availability for accomplishing the mission. Each subordinate maneuver element needs to understand their part of the mission, and leaders require awareness of any elements they have attached or detached for mission duration. Once leaders understand the mission, they then understand the equipment and personal protective equipment necessary for accomplishing the mission. When conducting mission analysis, leaders consider Soldier load management since excessive loads affect physical readiness and combat performance, and they can potentialize individual and small-unit mission failure. The acronym GEAR serves as a memory aid to leaders in managing Soldier load (see figure 2-1 and table 2-6).

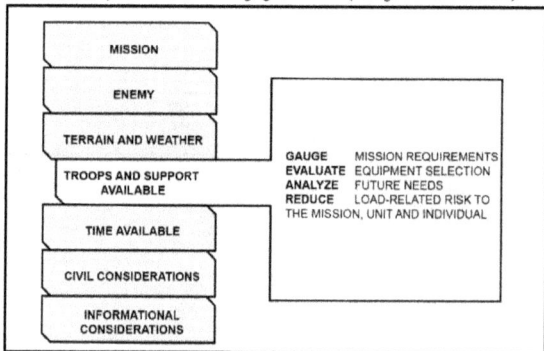

Figure 2-1. Gauge—evaluate—analyze—reduce framework

Table 2-6. Explanation of gauge–evaluate–analyze–reduce framework

Gauge mission requirements
Assess specific mission objectives for the individual or unit. Analyze environmental factors influencing equipment needs for the individual or unit. Analyze potential threats and challenges for the individual or unit.
Evaluate equipment options
Identify critical or essential equipment for mission success. Prioritize items based on mission critical tasks. Remember ounces mean pounds, and pounds impact performance. Ask, *What type of threat will the individual or unit face?* and   *What level of protection will the individual or unit need?*
Analyze future needs
Identify the most efficient, most effective way, depending on the mission variables, to carry the requisite load for mission success. Assess and coordinate off-loading platforms for equipment transportation and for individual and unit deployments. Coordinate for equipment and supply drops and caches as necessary and possible.
Reduce risk to the mission, individual, and unit

Analyze the risk to the mission introduced by Soldier load mismanagement in these two ways.
1) If Soldiers will be potentially carrying too much, identify whether extraneous equipment might—
  • Exceed a given Soldier's physical capacity.
  • Slow mission progress.
  • Lead to a failure to complete the mission within the given time constraints.
2) If Soldiers will be potentially carrying too little, identify whether exempted equipment might—
  • Introduce risk to the mission.
  • Negatively impact chances of mission success.
  • Negatively influence engagements with unexpected enemy threats.

(4) Time available. Leaders conduct deliberate time analysis and then disseminate a timeline to subordinates to ensure the mission is completed in accordance with the commander's timeline. Leaders conduct reverse planning from the not later than (NLT) time for mission completion in the higher's order to the time at which they will deliver the order. Doing so ensures their proper application of the 1/3::2/3 rule and affords subordinates enough time to prepare for the mission and to conduct rehearsals.

(5) Civil considerations. Analyze these. At the platoon level and below, they come largely from battalion-level analysis and the company commander's desired end state as regards the civilian populace and infrastructure. Are civilians present in the area of operations (AO)? Must civilian infrastructure be preserved? Is the support of the local populace critical for accomplishing follow-on operations? Leaders pay special attention to restrictions on indirect fire munitions.

STEPS 4–6—INITIATE MOVEMENT; CONDUCT RECONNAISSANCE; AND COMPLETE THE PLAN

2-6. The unit sometimes begins movement while the leader is still planning or forward reconnoitering. This step occurs anytime during the TLP.

2-7. When time allows, the leader may conduct personal reconnaissance. When time does not allow, the leader performs map reconnaissance. Sometimes, the leader relies on others such as scouts to conduct reconnaissance. The leader completes the plan based on reconnaissance and any changes in the situation.

STEP 7–ISSUE THE OPERATION ORDER

2-8. PLs and SLs normally issue oral OPORDs to aid subordinates in understanding the concept for the mission. Whenever possible, leaders issue the order with one or more of the following aids: within sight of the objective, on the defensive terrain, or on a terrain model or sketch.

2-9. Leaders may require subordinates to repeat all or part of the order or to demonstrate on the model or sketch their understanding of the operation. They also quiz their Rangers to ensure all Rangers understand the mission.

STEP 8–SUPERVISE AND REFINE

2-10. The leader supervises the unit's preparation for combat by conducting rehearsals and inspections. Rehearsals include the practice of having SLs brief their planned actions in execution sequence to the PL.

2-11. The leader conducts rehearsals on terrain resembling the actual ground and in similar light conditions. Leaders use rehearsals to—

- Have Rangers practice essential tasks, thereby improving performance.
- Reveal weaknesses or problems in the plan.
- Coordinate the actions of subordinate elements.
- Improve Rangers' understanding of the concept of operations, thereby fostering confidence.

2-12. The platoon sometimes begins rehearsals of battle drills and other SOP items before the receipt of the OPORD. After the issuance of the order, the platoon rehearses mission-specific tasks. Some important tasks to rehearse include the following:

- Actions on the objective.
- Actions at the assault position.
- Breaching obstacles such as mines and wire.
- Using special weapons or demolitions.
- Actions on unexpected enemy contact.

2-13. Different rehearsal types exist: confirmation brief, reduced force, full force, and techniques. During a confirmation brief rehearsal, key leaders sequentially brief the requisite actions for an operation. Patrol leader rehearsals are conducted twice—right after a FRAGORD (confirmation brief) and again after subordinates develop their own plans. A reduced force rehearsal is conducted when time is a key constraint and maintaining security is essential. Key leaders normally attend. Mockups, sand tables, and small-scale replicas are useful. A full force rehearsal is the most effective type. It is first executed in daylight and open terrain and then conducted in the same conditions as the operation. All Rangers participate and use force-on-force as appropriate.

2-14. The techniques rehearsal includes force-on-force, map (which has limited value and a limited number of attendees), radio (which cannot mass leaders but confirms communications), sand table or terrain model (which involves key leaders and includes all control measures), and the rehearsal of concept drill (which is similar to the sand table or terrain model except subordinates move themselves).

2-15. SLs conduct initial inspections shortly after receiving WARNORDs. The PSG spot-checks throughout the unit's combat preparation. The PL and PSG perform a final inspection. Precombat checks and inspections cover:

- Weapons and ammunition.
- Uniforms and equipment.
- Mission-essential equipment.
- Soldiers' understanding of the mission and individual responsibilities.
- Communications.
- Rations and water.
- Camouflage.
- Deficiencies noted during earlier inspections.

COMBAT INFORMATION

2-16. Gathering information is one of the most important aspects of conducting a patrolling operation. Report all information quickly, completely, and accurately. Use the SALUTE report format for recording and reporting information. (See table 2-7.)

Table 2-7. Example of a patrol report format

SIZE	Seven enemy personnel.
ACTIVITY	Traveling southwest.
LOCATION	GA123456.
UNIT / UNIFORM	Olive-drab uniforms with red six-point star on left shoulder.
TIME	210200 January 16.
EQUIPMENT	Carry one machine gun and one rocket launcher.

2-17. Try to include a field sketch with each report. Include only aspects of military importance such as targets, objectives, obstacles, sector limits, or troop dispositions and locations. Use symbols from FM 1-02.2. Use notes to explain the drawing without cluttering the sketch. Leave out information regarding personnel, weapons, and equipment; coverage of these items goes in the SALUTE report and not in this one.

2-18. The leader collects captured documents, submits them with the reports, and marks each document with the time and place of capture. When any prisoner captures occur during a patrolling operation, the treatment of those prisoners conforms to the Geneva Conventions and follows the 5-S rule. Immediately after returning from a mission, the unit undergoes debriefing. The 5 Ss of the 5-S rule are known to stand for search, silence, segregate, safeguard, and speed to rear.

2-19. A WARNORD gives subordinates advance notice of an upcoming operation, which, in turn, gives them time to prepare. A WARNORD is brief but complete. (See table 2-8 on pages 2-8 through 2-10 for a sample formatting of a WARNORD for a platoon. See table 2-9 on pages 2-10 through 2-15 for a sample formatting of a WARNORD for a squad.)

Note. A WARNORD authorizes execution only when it clearly says so.

Table 2-8. Example of a warning order format—platoon

WARNING ORDER (PLATOON)
Introduction and roll call; note-taking materials; RHB; map; protractor; a call to hold all questions until the end; leader(s) to monitor subordinates.
References: Refer to the higher HQ OPORD and identify map sheet for operation.
Time zone used throughout the order: Find this in the higher HQ OPORD.
Task organization: Brief on each squad's assigned role for the mission.
1. **SITUATION.** Find this in higher HQ OPORD para 1a (1–3). Include the following information.
  a. **Area of interest:** Outline the area of interest on the map.
    (1) Orient relative to each point on the compass (north, south, east, and west).
    (2) Box in the entire area with grid lines (see higher HQ OPORD para 1b).
  b. **Area of operations:** Outline the AO, using a map or terrain model. Point out the objective location and the current location of your unit.
    (1) Trace your zone boundaries; use features on the map that are generally linear in nature.
    (2) Familiarize your unit with their zone by identifying natural and artificial features. Brief on the significant deductions (conclusions) about each terrain feature from mission analysis.
    (3) Brief on high and low temperatures for the weather and the most significant weather factor.
  c. **Enemy forces:** Brief on who, what, and where (1 level up)—enemy forces (who), recent activities (what), and known and suspected locations with grid coordinates (where). Plot locations and then brief off the map board or map (see higher HQ OPORD para 1c [2]).
  d. **Friendly forces:** Brief on the commander's mission and intent and on adjacent units (1 level up).
    (1) State commander's mission (who, what, when, where, why) (see higher HQ OPORD para 2d).
    (2) State commander's intent with task and purpose (see higher HQ OPORD para 3a).
    (3) Point out friendly locations on the map board or map. State grid coordinates, task, and purpose.
  e. **Attachments and detachments.** (May include with task organization briefing.)
2. **MISSION.** Provide the mission statement twice (who, what, when, where, why). For example,
1st Platoon, B Company decisive operation attacks (OFFENSIVE TASK) to destroy (EFFECTS ON ENY) enemy personnel and equipment on OBJ Red (GA152793) NLT 302300 January 17 in order to prevent the enemy from maintaining control of OBJ Red (PURPOSE).

Table 2-8. Example of a warning order format—platoon (continued)

3. **EXECUTION.** Include the following information at a minimum.
  a. **Concept of operations:** Describe the execution of your unit's mission in general terms
    from start to finish. Cover all major movements. Follow the KDDTMK format—key action,
    distance, direction, time of movement, method of movement, key location (grid coordinates for
    checkpoints with key terrain).
  b. **Tasks to subordinate units:** Provide specific tasks to subordinate units to aid in the planning,
    preparation, and execution of the mission. Planning guidance—
    (1) Includes tasks assigned to subordinates and subordinate leaders.
    (2) Is issued with T/C/S and a TH.
    (3) Allows subordinates to execute the task with minimal ongoing supervision.
  c. **Coordinating instructions:** Include any known information currently available. Cover the
    uniform and equipment common to all. Consider the mission variables when tailoring the
    Soldier load to meet mission-specific requirements in accordance with the GEAR framework.
    Regarding the timeline, state who, what, when, and where and any relevant notes. (All specific
    times must be listed.) Give specific priorities in order of completion, information about
    coordination meetings, rehearsals, and inspections by priority, and the NLT time to start
    movement.

4. **SUSTAINMENT.** Include known logistics preparations for the operation.
  a. **Logistics:** Include the following information.
    (1) **Maintenance:** Include weapons and equipment DX time and location.
    (2) **Transportation:** State method and mode of transportation for INFIL and EXFIL. Identify the
    necessary coordination for external assets. Task subordinate leaders as necessary with
    generating load plan, number of lifts or serials, and bump plan.
    (3) **Supply:** Only include classes of supply that require coordination or special instructions.
    Break down the information by supply class (relevance to mission), on-hand quantity,
    requested quantity, and distribution method. For resupply, include time and location.
  b. **Army Health System support:** Identify the medical equipment, medical support, and preventive
    medicine that require coordination.

5. **COMMAND AND SIGNAL.**
  a. **Command:** State the succession of command (top four) if it is not covered in the unit's SOP.
  b. **Control:** Include the following information.
    (1) **CPs:** Describe the employment of CPs, including the location and the opening and closing
    times of each CP as appropriate. (Typically, at the platoon level, the only reference to CPs
    is to the company CP.)
    (2) **Reports:** List reports not covered in the unit's SOP (OPSKED).
  c. **Signal:** Describe the concept of signal support, including current SOI, or refer to higher HQ
    OPORD para 5.

Table 2-8. Example of a warning order format—platoon (continued)

Additional Guidance. 1. State next specified time and required key actions. 2. Ask for questions. 3. Conduct backbriefs as time permits. 4. Give time, place, and uniform of the OPORD. 5. Give TH (time synchronization).
Legend: AO–area of operations; CP–command post; DX–direct exchange; ENY–enemy; EXFIL–exfiltration; GEAR–gauge, evaluate, analyze, reduce; HQ–headquarters; INFIL–infiltration; NLT–not later than; OBJ–objective; OPORD–operation order; para–paragraph; OPSKED–operation schedule; RHB–*Ranger Handbook*; SOI–signal operating instructions; SOP–standard operating procedure; T/C/S–tasks, conditions, standards; TH–time hack

Table 2-9. Example of a warning order format—squad

WARNING ORDER (SQUAD) Introduction and roll call; CDS; note-taking materials; RHB; 1:50,000 map; protractor; call to hold all questions until the end; TLs to monitor task organization. **ATM–SEC / BTM–ASSLT / HQ SUPPORT.**

1. **SITUATION.** Brief.
  a. **Area of interest:** Orient the map (north, south, east, and west). For example, **our SQD's area of interest is boxed in by the 86 grid line to the north, the 18 grid line to the east, the 77 grid line to the south, and the 13 grid line to the west.**
  b. **Area of operations:** For example, **we will be operating in Zone C.** (Trace Zone C with boundaries. Contextualize Zone C with three natural and artificial features.) **Our OBJ is located here** (Indicate on map.) **at GA152796, and our current location is here** (Indicate on map.) **at GA196790.**
  c. **Enemy:** Use the 3 Ws and para 1c (1–3). Describe the enemy's recent locations and activities.
    (1) **WHO?** The Aragon Liberation Front.
    (2) **WHAT?** Ambushed patrol.
    (3) **WHERE?** GA156804.
  d. **Friendly:** Use the 4 Ws, para 2 (Mission), intent, and concept one and two levels up. Task and purpose of adjacent patrols. Provide the big picture concept of <u>higher HQ's mission and intent.</u>
    (1) **Mission.**
      (a) **WHO?** 1st PLT, B CO.
      (b) **WHAT?** (Task) Conduct area ambushes to destroy enemy forces.
      (c) **WHERE?** On OBJ Black NLT 302300 January 17.
      (d) **WHY?** (Purpose) To prevent the enemy from maintaining control of OBJ Black.
    (2) **Intent.**
      (a) Find, fix, and finish enemy forces in Zone C.
      (b) Destroy enemy personnel and equipment
  e. **Attachments and detachments:** MG TM 300530 January 17.

Table 2-9. Example of a warning order format—squad (continued)

2. **MISSION.** Use clear and concise language, the 5 Ws, para 3, X2, and task and purpose. For example, **1st SQD, 1st PLT, B CO decisive operation attacks (OFFENSIVE TASK) to destroy (EFFECTS ON ENY) enemy personnel and equipment on OBJ Red (GA152793) NLT 302300 January 17 in order to prevent the enemy from maintaining control of OBJ Red (PURPOSE).**

3. EXECUTION.
   a. **Concept of operations:** (Orient Rangers to sketch or terrain model.) We are currently located at Camp Darby, GA19627902. We will depart Camp Darby, moving generally northwest for 6,000 meters. We will travel by truck to our insertion point, GA176812, where we will dismount the trucks. The movement should take approximately 20 minutes. Our ground tactical plan will begin when we move generally southwest for 3,000 meters. The movement should take 3 hours as we travel by foot to our tentative ORP, GA154795. Here. we will finalize personnel weapons, and equipment preparations. We will then move generally southwest for 400 meters to our OBJ, GA152793. The movement by foot should take 30 minutes to an hour. From our OBJ, we will travel generally northwest for 3,000 meters. The movement by foot to our linkup site, GA152819, should take 4 hours. Once linked up. we will move generally southeast for 8,000 meters. The movement should take approximately 30 minutes as we travel by truck back to Camp Derby. Once back, we will debrief and prepare to conduct follow-on operations.
   b. **Tasks to subordinate units:** Nontactical and tactical instructions, METT-TC (I) planning guidance, TMs, special TMs, and key individuals (control–movement–objective).
      (1) **HQ:** Second in the OOM, the M240 provides supporting fire into the kill zone during AOO. The RTO is the recorder en route and during AOO. You write para 5 of the SQD OPORD, ensure all radios are operational with proper frequencies loaded, and ensure we enter the net on time.
      (2) **ATM:** First in the OOM, the ATM is responsible for land navigation. The ATM is flank security for AOO, one- to two-Ranger EPW TM, one- to two-Ranger aid and litter TM, one- to two-Ranger demo TM, one- to two-Ranger ORP-clearing TM, two two-person Ranger flank security TMs for AOO, one- to two-Ranger linkup security TM, one SQD automatic weapon gunner ASSLT element for AOO, one compass person, and one pace person. The ATL is the security TL for AOO. You write para 1 and the linkup annex of the SQD OPORD and draw all sketches. You are responsible for the formation order of movement, danger areas, TMs, special TMs, battle drills, linkup, truck, AOO, terrain model, routes, and fire support overlay (sterile and nonsterile).
      (3) **BTM:** Third in the OOM, the BTM is ASSLT for AOO. One- to two-person EPW TM. one- to two-Ranger aid and litter TM, one- to two-Ranger demo TM, one- to two-Ranger surveillance and observation TM, one grenadier per security TM for AOO, one compass person, and one pace person. The BTL is the ASSLT TL for AOO. You are second in the COC and in charge at all times during my absence. You write para 4 and the truck annex of the SQD OPORD, prepare the supply, DX, and ammunition lists, and draw and issue all items. Ensure everyone does a T/F and that all equipment is tied down in accordance with the 4th Ranger Training Battalion's SOP. Update the SQD status card and hand receipt.

Table 2-9. Example of a warning order format—squad (continued)

3. **EXECUTION** (continued).
    (4) **TL:** The TL updates the WARNORD board with all the correct information. As a task is accomplished, you draw a line through it. You post your COC, duty position, and job description (special TMs and key individuals). You come see me for further guidance at the conclusion of this WARNORD.
  c. **Coordinating instructions:** Depending on METT-TC (I), tailor the load appropriately to the number of Rangers and not in accordance with the SOPs.
    (1) The packing list is based on the Airborne and Ranger Training Brigade's seasonal packing list.
    (2) Write on note cards or paper and read off by item.

WORN

PACKING LIST
ON NOTE CARD

RUCK

PACKING LIST
ON NOTE CARD

Time schedule

WHEN	WHAT	WHERE	WHO	REFERENCE
*0630	WARNORD	Bay area	All	3D1
0700	Initial inspection	Bay area	All	
*0730	Required ammo, supply	Co TOC	BTL	4B1
*0745	Pick up ammo, supply	Co TOC	BTL/Detail	4B1
0800	T/F	T/F area	All	
*0830	S–2, S–3, fires coordination	PLT bay	SL/RTO	3D1
0850	Adjacent unit coordination	SQD bays	ATL, Compass	3D1
*0900	Enter net	Bay area	RTO	
0930	Status card update	Bay area	SL/TL	RTO
0945	Terrain model complete	Bay area	SL/TL	
*1000	OPORD	Bay area	All	

Table 2-9. Example of a warning order format—squad (continued)

Time schedule (continued)				
WHEN	WHAT	WHERE	WHO	REFERENCE
1300	Rehearsal	Bay area	All	
1330	Final inspection	Bay area	All	
*1400	Truck linkup	Co TOC	All	3D1
1500	Depart Camp Darby	Co TOC	All	
*1600	Insertion complete	TBD	All	3D1
1800	In ORP	TBD	All	
*2000	In position	GA184774	All	
*2300	Mission complete	TBD	All	
*0200	Linkup complete	TBD	All	
0500	S–2 debriefing	Battalion TOC	All	
* Critical times. Use the 1/3::2/3 rule and reverse planning.				

4. **SUSTAINMENT.** Include the following information.
   a. **Logistics.**
     (1) **Maintenance:** Weapons and equipment DX is at 0700 in the company CP.
     (2) **Transportation:** Trucking will be the mode of transportation for insertion and extraction. The BTL generates the load plan, bump plan, and number of chalks and lifts.
     (3) **Supply.**
       (a) Class I—food, rations, and water: Each person has two MREs and 6 quarts of water for the operation.
       (b) Class V—ammunition: The BTL draws enough ammunition for each person to carry a basic load in accordance with the SQD's SOP.

Table 2-9. Example of a warning order format—squad (continued)

4. SUSTAINMENT (continued).

Example of a squad ammunition standard operating procedure

INDIVIDUAL WEAPONS (FOR EACH WEAPON SYSTEM)		
WPN	QTY	REMARKS
M4	210	5.56-mm ball
M249	625	5.56-mm ball, link (4:1 mix)
M240B	825	7.62-mm ball, link (4:1 mix)
M320	12	40-mm HEDP
M320	3	40-mm shot
M320	4	40-mm illumination
M320	2	40-mm TP rounds

SUPPLEMENTAL AMMUNITION (TOTAL)		
TYPE	QTY	REMARKS
Claymore	3	M18A1
Demo kit	2	C4, M81, M14
M67 hand grenade	24	Two per Soldier
5.56-mm tracer	126	42 per leader
AT4 (antitank)	2	One per TM
HC smoke	6	
Red smoke	2	
Yellow smoke	4	

b. **Personnel services support:** Religious services will be held at 0800 in the chapel.

c. **Army Health System support:** The ATL coordinates for one additional combat lifesaver bag.

5. COMMAND AND SIGNAL. Include the following information.

 a. **Command.**

  (1) **Location of commander or patrol leader:** The patrol leader will be located in the SQD bay during phase one (mission preparation). The OPORD briefs the location of the patrol leader for all other phases.

  (2) **Succession of command:** State the succession of command when the unit's SOP does not. This is the SL, BTL, ATL, and RTO.

 b. **Control.**

  (1) **CPs:** The PLT CP is located at GA166807, and the company CP is located at GA196790.

  (2) **Reports:** The OPORD covers pertinent reports.

 c. **Signal.**

  (1) The battalion will be operating on 37.950 SC/PT, call sign Darby 741.

  (2) Our SQD frequency is 77.000 SC/PT, call sign bravo-one-one.

  (3) The OPORD briefing gives all other signals such as frequencies, call signs, challenges, and passwords.

Table 2-9. Example of a warning order format—squad (continued)

Additional Guidance. 1. Give subordinates additional guidance on tasks to complete for preparation of the OPORD and mission. 2. Give the time, place, and uniform of the OPORD. 3. Give a time hack and ask for questions. 4. Time permitting, complete a subordinate confirmation brief on the distributed information to ensure understanding.
Legend: 3 Ws–who, what, where; 4 Ws–who, what, where, why; 5 Ws–who, what, where, when, why; ammo–ammunition; AOO–actions on objective; ASSLT–assault; ATL–Alpha team leader; ATM–Alpha team; BTL–Bravo team leader; BTM–Bravo team; CDS–Camp Darby special; CO–company; COC–chain of command; CP–command post; demo–demolition; DX–direct exchange; ENY–enemy; EPW–enemy prisoner of war; HC–high concentration; HEDP–high explosive dual purpose; HQ–headquarters; METT-TC (I)–mission, enemy, terrain and weather, troops and support available, time available, civil considerations, and informational considerations; MG–machine gun; mm–millimeter; MRE–meal, ready to eat; NLT–not later than; OBJ–objective; OOM–order of march; OPORD–operation order; ORP–objective rally point; para–paragraph; PLT–platoon; QTY–quantity; RHB–*Ranger Handbook*; RTO–radio-telephone operator; S–2–battalion or brigade intelligence staff officer; S–3–battalion or brigade operations staff officer; SC/PT–single channel/plain text; SEC–security; SL–squad leader; SOP–standard operating procedure; SQD–squad; TBD–to be determined; T/F–test fire; TL–team leader; TM–team; TOC–tactical operations center; TP–training practice; WARNORD–warning order; WPN–weapon; X2–times two

OPERATION ORDER

2-20. An OPORD is a directive issued by a leader to subordinates in order to effect the coordinated execution of a specific operation. A five-paragraph format (see table 2-10 on pages 2-16 through 2-23) organizes the briefing, ensures completeness, and helps subordinate leaders understand and follow the order. Use a terrain model or sketch, along with a map, to explain the order.

2-21. The PL or SL orally briefs the OPORD off notes. Before the issuance of the OPORD, the leader ensures note-taking materials, copy of *Ranger Handbook*, map, and protractor are all in place. The leader monitors subordinates, calls roll, and says PLEASE HOLD ALL QUESTIONS UNTIL THE END. Table 2-10 is a sample formatting of an OPORD for a platoon.

Table 2-10. Example of an operation order format—platoon

OPERATION ORDER. Plans and orders normally contain a code name and are numbered consecutively within a calendar year.

References: The heading of the plan or order lists maps, charts, data, or other documents the unit needs to understand the plan or order. The user may reference the SOP in the body of the plan or order. Map references are by map series number and sheet number (and, if required, by country or geographic area, name, edition, and scale). The term *datum* refers to the orientation of a map's lines of latitude and longitude relative to the earth. Different nations use different datums for printing coordinates on their maps. Knowing the datum in use helps when determining coordinates; look first in a map's marginal information, which usually references the datum. Include the following information.

Time zone used throughout the order: For an operation occurring in one time zone, use that time zone throughout the order, including in the annexes and appendixes. For an operation spanning several time zones, use Zulu time.

Task organization: Describe the allocation of forces to support the commander's concept. Show task organization in one of two places: just above paragraph 1 or, for a long or complex task organization, in an annex. Provide a task and purpose for each subordinate maneuver element and designate a decisive operation. The decisive operation's purpose nests within the platoon's purpose.

1. **SITUATION.** Include the following information.
   a. **Area of interest:** Describe areas of interest outside the AO that influence your operation.
      (1) Orient the audience to the map or visual aid. Show north, south, east, and west.
      (2) Box in the area of interest with grid lines. Horizontal grid lines indicate northing, and vertical grid lines indicate easting.
      (3) Brief on enemy external assets in the area of interest. Follow the CARRO format—
         (a) **Close air support:** Enemy fixed- or rotary-wing assets (available for use by the enemy).
         (b) **Artillery/mortars:** Enemy indirect fire assets that can reach friendly forces at any point during the operation. Pay close attention to munitions and maximum effective ranges.
         (c) **Reinforcements:** Enemy quick reaction forces who can support enemy forces within the AO. How quickly can the enemy be reinforced in the AO?
         (d) **Reserves:** Normally controlled at a higher echelon. What forces does the enemy have in reserve?
         (e) **Other:** What other assets does the enemy have that can affect friendly forces?

Table 2-10. Example of an operation order format—platoon (continued)

1. **SITUATION** (continued).
   b. **Area of operations:** Describe the AO. Refer to the appropriate map and use overlays as necessary. Trace the platoon's AO, basing the boundaries on visually recognizable natural and artificial features on the map. Familiarize the audience with the AO by identifying three artificial and three natural features. Since the subsequent paragraph provides a full terrain analysis, no deductions (conclusions) about the terrain features need be included here (as a warning order briefing would include).
      (1) **Terrain:** Use the OAKOC from the higher HQ OPORD. Refine it based on your analysis of the terrain in the AO. Using the OAKOC format, state how the terrain will affect friendly and enemy forces in the AO. Provide deductions (conclusions) about each OAKOC military aspect of terrain.
      (2) **Weather:** Describe the aspects of weather that impact operations. Allow the military aspects of weather to drive your analysis—visibility, wind, precipitation, cloud cover, temperature, humidity, and atmospheric pressure. State how the weather will affect both friendly and enemy forces in the AO. Provide deductions (conclusions) about each weather condition.
   c. **Enemy forces:** The enemy situation in higher HQ OPORD para 1c forms the basis for this. Refine it by adding the details your subordinates require.
      (1) Brief on and depict enemy forces two and one level(s) up, as stated in the company order.
      (2) Regarding the enemy forces templated in your AO—
         (a) State the enemy's composition and strength.
         (b) Brief on and depict the enemy's disposition along the avenues of approach in your AO.
         (c) Brief on and depict the enemy's recent activities.
         (d) Describe enemy capabilities, focusing on direct, indirect, and reinforcement capabilities.
         (e) Describe the enemy's MPCOA and MDCOA. Focus the MPCOA and MDCOA on the AOO. Brief off the micro-terrain model.
   d. **Friendly forces:** Find this in higher HQ OPORD paras 1d, 2. and 3.
      (1) **Higher HQ mission, intent, and concept.**
         (a) Higher HQ two levels up.
            _1_ Mission: State the mission of the higher unit two levels up.
            _2_ Intent: State the intent two levels up.
         (b) Higher HQ one level up.
            _1_ Mission: State the mission of the higher unit one level up.
            _2_ Intent: State the intent one level up.
      (2) **Mission of adjacent units:** State the locations of units to the left, right, front, and rear; those units' tasks and purposes; and how those units will influence yours, particularly adjacent unit patrols. Conduct this analysis for friendly forces one and two level(s) up.
         (a) Show other units' locations on the map board.
         (b) Include statements about what, if any, influence each patrol will have on your mission.
         (c) Find this information in company order paras 1d and 3. It gives each leader an idea of what other units are doing and where they are going.
         (d) Include any information obtained when the leader conducts adjacent unit coordination.

Table 2-10. Example of an operation order format—platoon (continued)

1. **SITUATION** (continued).

   e. **Attachments and detachments:** Avoid repeating information already listed in task organization, but list those units attached or detached to the HQ issuing the order when not listed in task organization. State when attachment or detachment will be in effect if different from when the OPORD is in effect, such as on order or on commitment of the reserve. Use the phrase *remains attached* when units will be or have been attached for some time.

2. **MISSION.** Who, what (task), when, where, why (purpose) from higher HQ OPORD Maneuver para.

3. **EXECUTION.**

   a. **Commander's intent:** State the intent, which is the clear, concise statement of what the force must do and of the conditions the force must establish (with respect to the friendly, enemy, terrain, and civil considerations) to achieve the commander's desired end state. This statement allows subordinate and supporting commanders to execute the mission without further orders, even when the operation does not unfold as planned. Do not repeating when already briefed off para 1d.

   b. **Concept of operations:**

      (1) Write a clear, concise course of action statement.

         (a) State the form of maneuver to be used to accomplish the mission.
         (b) State the decisive point of the operation and explain what makes it decisive.
         (c) Brief on the task and purpose of subordinate maneuver elements in the operation.
         (d) State the purpose of the enablers throughout the operation. Separate the enablers by warfighting function.
         (e) Brief on the commander's desired end state for the friendly, enemy, terrain and civil variables.

      (2) Brief on the nested concept by phase. Describe how the unit will accomplish its mission from start to finish. Base the number of subparagraphs, if any, on what the leader considers appropriate, the level of leadership, and the complexity of the operation. At a minimum, divide the operation into the following phases: planning and preparation, movement, AOO, and preparation for follow-on operations. State when each phase will begin and end and what is critical to success during the phase.

   c. **Scheme of movement and maneuver:** The organization of this brief is sequential, moving from the start of the operation to its finish and covering all aspects of the operation. Describe the operation from when the platoon begins movement to when the operation is complete.

      (1) Brief off the macro-terrain model on the primary and alternate routes. Follow the KDDTMK format—key action, distance, direction, time of movement, method of movement, key location (key terrain at checkpoint). State the movement technique and formation. Provide a grid for the endpoint of each leg of the movement.

      (2) Cover the primary and alternate routes from insertion or start point, through AOO, and to linkup or arrival at the subsequent patrol base.

      (3) Brief on the plan for crossing known dangerous areas. Provide near side and far side rally points and tentative crossing points.

Table 2-10. Example of an operation order format—platoon (continued)

3. **EXECUTION** (continued).
    (4) Brief on the plan for reacting to enemy contact. Brief on each contingency to the leg of the movement that you most anticipate taking contact.
        (a) For direct contact, consider the following.
          _1_ When contact is made, which element would be most likely to take contact, and based on your analysis, from which direction is direct fire contact most likely?
          _2_ How will you assess the situation? Applying three-to-one superiority, what sizes of enemy elements and weapon systems by type can be engaged? How does the terrain affect maneuverability? What are the criteria for employing the key weapon systems? What is the decision point for conducting a platoon attack or breaking contact?
          _3_ Based on your analysis, how will you task-organize the platoon? Which elements will serve as the base-of-fire, the maneuver, and the security elements?
          _4_ What control measures will you implement for the course of action on which you will brief (for example, signals to cease or shift fire)?
          _5_ When anticipating conducting a platoon attack, how will you employ special teams to exploit the enemy?
        (b) For indirect contact, consider the following.
          _1_ Upon receiving indirect fire on a leg of the movement, in which direction and to what distance will the movement travel?
          _2_ What is to be the reaction upon receiving bracketing fire?
        (c) Brief on any other likely forms of contact during the patrol's legs of the movement.
    (5) Brief on any approved fire targets, CCPs, CXPs, and HLZs as they become applicable throughout the route.
    (6) Brief on the scheme of maneuver chronologically within the scheme of movement. Upon completing the scheme of maneuver, continue through the scheme of movement until mission completion. Upon completing the primary route, brief on the alternate route in the KDDTMK format. The scheme of maneuver includes the following at minimum.
        (a) **Movement plan:** How elements will move to and from the objective.
        (b) **Reconnaissance plan:** How leaders will reconnoiter positions on the objective and who will accompany them.
        (c) **Occupation plan:** How leaders will move Rangers into positions on the objective.
        (d) **Engagement/disengagement criteria:** A brief based on conditions on the objective such as time, emplaced personnel, and enemy. Rangers maintain understanding of when and how they are to engage the enemy on the objective (for example, priority of targets, rates of fire by weapon system).
        (e) **Compromise plan:** Actions each subordinate element will take upon hard or soft compromise.
        (f) **Control measures:** Which control measures—signals, target reference points, limit of advance, land component commander, restrictive fire line—to implement to maximize effectiveness and mitigate fratricide.

Table 2-10. Example of an operation order format—platoon (continued)

3. **EXECUTION** (continued).

      (g) **Initiation plan:** How the platoon will initiate AOO. A PACE plan is recommended to ensure complete redundancy.

      (h) **AOO plan:** How the platoon will conduct the operation upon initiation. How the element will fire and maneuver. How special teams will be employed to exploit the objective.

      (i) **Withdrawal plan:** How the platoon will withdraw from the objective once AOO are complete.

      (j) **Consolidation and reorganization:** How the platoon will consolidate and reorganize for follow-on operations once AOO are complete.

  d. **Scheme of fires:** State the scheme of fires to support the overall concept and state who (which maneuver unit) has priority of fires by the phase of the operation. Discuss specific fire targets by the TTLODAC format and point to them on the macro-terrain model (see chapter 3).

  e. **CASEVAC:** Brief on the locations of CCPs, CXPs, and HLZs and indicate them on the macro-terrain model. State when they will be active during the operation.

  f. **Tasks to subordinate units:** Task Rangers in subordinate squads with special teams—EPW, aid and litter, vehicle clearance, demolitions, surveillance and observation, objective rally point security. Brief on what special equipment these personnel need and when they will conduct rehearsals. Place into the coordinating instructions those tasks that affect two or more units.

  g. **Coordinating instructions:** This is always the last subparagraph under paragraph 3 (Execution). List only those instructions that apply to two or more units. Use annexes as necessary for communicating more complex instructions.

    (1) **Timeline:** State the time, place, uniform, and audience for each key event. Cover the timeline from the briefing of the order through mission completion. Brief on the priority of rehearsals, confirmation briefs, and inspections.

    (2) **Commander's critical information requirements:** Include PIRs and FFIRs.

      (a) **PIRs:** The necessary intelligence for the commander and staff to understand the threat and other aspects of the operational environment.

      (b) **FFIRs:** The necessary information for the commander and staff to understand the status of friendly force and supporting capabilities. This may include personnel status, ammunition status, and leadership capabilities.

    (3) **Essential elements of friendly information:** Critical aspects of a friendly operation that compromise or limit the success of the operation or lead to its failure when the enemy knows them. Therefore, they are protected from enemy detection.

    (4) **Risk-reduction control measures:** Unique to the operation, these supplement the unit's SOP and include mission-oriented protective posture, operational exposure guidance, vehicle recognition signals, and fratricide prevention measures.

    (5) **Rules of engagement.**

    (6) **Environmental considerations.**

    (7) **Force protection.**

Table 2-10. Example of an operation order format—platoon (continued)

4. **SUSTAINMENT.** Describe the concept of sustainment to include logistics, personnel, and medical.
   a. **Logistics:** Include the following information.
      (1) **Sustainment overlay:** Include current and proposed company trains locations, CCPs, equipment collection points, HLZs, CXPs, EPW exchange points, and any friendly sustainment locations such as forward operating bases. Include grid locations. Brief on the locations with the use of a map or terrain model.
      (2) **Maintenance:** Include weapons and equipment, direct exchange time. and location. Include the weapons maintenance plan for after AOO are complete.
      (3) **Transportation:** State the method and mode of transportation for insertion and extraction, load plan, number of lifts and serials, bump plan, recovery assets, and recovery plan. When vehicles will provide a method of transportation, include a plan for security at the insertion and extraction points.
      (4) **Supply:** Include the current status of the supplies.
         (a) Class I—food, rations, and water.
         (b) Class III—petroleum, oils. and lubricants.
         (c) Class V—ammunition and explosives.
         (d) Class VII—major end items.
         (e) Class VIII—medical materiel.
         (f) Class IX—repair parts.
         (g) Distribution methods.
      (5) **Emergency resupply plan:** Brief on an emergency resupply plan for Class I. V, and VII supplies. Break down each class into materials within that supply class (for example. separate Class I into food and water). Brief in the following order.
         (a) Class of supply and the specific material.
         (b) On-hand quantity that would trigger an emergency resupply. The platoon sergeant maintains awareness of what on-hand quantity of each key resource the platoon must have to remain combat effective.
         (c) Quantity to request for each resource in an emergency resupply.
         (d) How each resource should be packaged.
         (e) How to mark each arriving package.
         (f) The necessity of external assets to provide security for the package, as well as what size security detail to request when necessary.
      (6) **Field services:** Include any services provided or required.
   b. **Personnel services support:** Include the methods of marking and handling EPWs.
      (1) Brief on how to handle EPWs in accordance with search. silence, segregate, safeguard, and speed to the rear.
      (2) Identify EPW teams by name and state any special equipment they are to carry.
      (3) Provide a grid for each EPW collection point and indicate them on the macro-terrain model, but not when doing so would be redundant to the sustainment overlay briefing.
      (4) Brief on an EPW backhaul SOP. Using a visual aid, provide a detailed formation and order of movement for transporting EPWs.

Table 2-10. Example of an operation order format—platoon (continued)

4. **SUSTAINMENT** (continued).

    c. **Army Health System support:** Include the following information.

        (1) **Medical mission command:** Include the locations of the medic and the platoon sergeant.

        (2) **Medical treatment:** State how wounded or injured Soldiers will receive treatment (for example, self-aid, buddy aid, combat lifesaver bag, emergency medical technician).

        (3) **Medical evacuation:** Identify aid and litter teams by name and state any special equipment they are to carry.

        (4) **Preventive medicine:** Identify any preventive medicine needs for the mission (for example, sun block, lip balm, insect repellant, in-country-specific medicine).

    d. **CASEVAC plan:** Brief on a detailed CASEVAC plan. Use visual aids for the following.

        (1) CCP establishment and marking methods for both good and limited visibility.

        (2) CASEVAC formation and order of movement. Brief on where each squad will be while transporting casualties. The formation is not personalized down to the individual.

        (3) CXP establishment and marking methods for both good and limited visibility.

        (4) A detailed mass casualty plan on the objective, specifically using a micro-terrain model.

            (a) Include a trigger—that is, how many casualties by type constitute a mass casualty. Include how the platoon will receive an alert to the mass casualty situation.

            (b) Discuss the location of your CCP and task-organize the platoon to secure it.

            (c) Brief on how your CCP will triage casualties. A CCP diagram is helpful for demonstrating the planned conduct of patient triage.

5. **COMMAND AND SIGNAL.** State the locations of mission command facilities and key leaders for the operation.

    a. **Command:** Include the following information.

        (1) **Location of commander or patrol leader:** State where the commander intends to be during the operation, by phase whenever the operation is phased.

        (2) **Succession of command:** State the succession of command when the unit's SOP does not.

    b. **Control:** Include the following information.

        (1) **CPs:** Describe the employment of any CPs, including the location of each CP and its times of opening and closing. Typically, at platoon level, any reference to CPs is to the company CP.

        (2) **Reports:** List reports not covered in the SOPs.

    c. **Signal:** Describe the concept of signal support, including current SOI edition, or refer to the higher HQ OPORD.

        (1) Identify the **SOI index** in effect.

        (2) Identify **methods of communication by priority.**

        (3) Describe **pyrotechnics and signals** such as hand and arm and demonstrate as appropriate.

        (4) Give **code words** such as OPSKEDs.

        (5) Give the **challenge and password** for use behind friendly lines.

        (6) Give the **number combination** for use forward of friendly lines.

        (7) Give the **running password.**

        (8) Give **recognition signals** (near—far, day—night).

Table 2-10. Example of an operation order format—platoon (continued)

Actions after the issuance of the OPORD: 1. Issue annexes. 2. Highlight next hard time. 3. Give time hack. 4. Ask for questions.
Legend: AO–area of operations; AOO–actions on objective; CARRO–close air support, artillery/mortars, reinforcements, reserves, other; CASEVAC–casualty evacuation; CCP–casualty collection point; CP–command post; CXP–casualty exchange point; EPW–enemy prisoner of war; FFIR–friendly force information requirement; HLZ–helicopter landing zone; HQ–headquarters; KDDTMK–key action, distance, direction, time of movement, method of movement, key location; MDCOA–most dangerous course of action; MPCOA–most probable course of action; OAKOC–observation and fields of fire, avenues of approach, key terrain, obstacles, and cover and concealment; OPORD–operation order; OPSKED–operation schedule; PACE–primary, alternate, contingency, emergency; para–paragraph; PIR–priority intelligence requirement; SOI–signal operating instructions; SOP–standard operating procedure; TTLODAC–target, trigger, location, observer, delivery system, attack guidance, communications network

FRAGMENTARY ORDER

2-22. A FRAGORD is an abbreviated form of an OPORD, usually issued daily, which eliminates the need for restating portions of the OPORD. Its issuance follows an OPORD to change or modify that order or to execute a branch or sequel to that order. (See table 2-11 on pages 2-23 through 2-26 for a sample formatting of a FRAGORD with annotations.)

Table 2-11. Example of a fragmentary order format

FRAGMENTARY ORDER
Time zone referenced throughout order:
Task organization:
1. **SITUATION.** Brief changes from the base OPORD specific to the day's operation. Include the following information.
  a. **Area of interest:** State any changes to the area of interest.
  b. **Area of operations:** State any changes to the area of operations. These include terrain—observation, fields of fire, cover and concealment, obstacles, key terrain, and avenues of approach. Note any changes that will affect the operation in the new area of operations.

Table 2-11. Example of a fragmentary order format (continued)

1. **SITUATION** (continued).
  c. **Enemy:** Include the following information.
    (1) Composition, disposition, and
      strength.
    (2) Capabilities.
    (3) Recent activities.
    (4) Most likely course of action.
  d. **Friendly:** This includes:
    (1) Higher mission.
    (2) Adjacent patrols, tasks, and
      purposes.
    (3) Adjacent patrol's objective and
      route if known.

Temp high	Sunrise	Moonrise
Temp low	Sunset	Moonset
Wind speed	*BMNT	Moon phase
Wind direction	**EENT	Percent illumination

This is the information the squad leader received from the platoon OPORD.
*Begin morning nautical twilight
**End of evening nautical twilight

2. **MISSION.** Who, what (task), when, where, and why (purpose) from the higher HQ OPORD Maneuver para.

3. **EXECUTION.** Include the following information.
  a. **Commander's intent:** Include any changes or state NO CHANGE.
  b. **Concept of operations:** Include any changes or state NO CHANGE.
  c. **Scheme of movement and maneuver:** Include any changes or state NO CHANGE.
  d. **Scheme of fires:** Include any changes or state NO CHANGE.
  e. **Casualty evacuation:** Include any changes or state NO CHANGE.
  f. **Tasks to subordinate units:** Include any changes or state NO CHANGE.
  g. **Coordinating instructions:** Include any changes or state NO CHANGE. Add the following information.
    (1) Time schedule.
    (2) Commander's critical information requirement(s).
    (3) Priority intelligence requirement(s).
    (4) Friendly force information requirement(s).
    (5) Essential elements of friendly information.
    (6) Risk reduction control measures.
    (7) Rules of engagement.
    (8) Environmental considerations.
    (9) Force protection.

4. **SUSTAINMENT.** Cover only the changes from the base order. Use the standard format. Brief items that have not changed as NO CHANGE.
  a. **Logistics:** Include the following information.
    (1) **Sustainment overlay.**
    (2) **Maintenance.**
    (3) **Transportation.**

Table 2-11. Example of a fragmentary order format (continued)

4. **SUSTAINMENT** (continued).
    (4) **Supply.**
        (a) Class I—food, rations, and water.
        (b) Class III—petroleum, oils, and lubricants.
        (c) Class V—ammunition and explosives.
        (d) Class VII—major end items.
        (e) Class VIII—medical materiel.
        (f) Class IX—repair parts.
        (g) Distribution methods.
    (5) **Field services.**
  b. **Personnel services support:** Include the following points.
    (1) Methods of marking and handling enemy prisoners of war.
    (2) Religious services.
  c. **Army Health System support:** Include the following points.
    (1) Medical mission command.
    (2) Medical treatment.
    (3) Medical evacuation.
    (4) Preventive medicine.

5. **COMMAND AND SIGNAL.** Brief only the changes to the base order. With changes, state the locations of mission command facilities and key leaders for the operation.
  a. **Command:** Include the following information.
    (1) **Location of commander or patrol leader:** State where the commander intends to be during the operation, by phase whenever the operation is phased.
    (2) **Succession of command:** State the succession of command when the unit's SOP does not.
  b. **Control:** Include the following information.
    (1) **CPs:** Describe the employment of any CPs including the location of each CP and its times of opening and closing. Typically, at platoon level, any reference to CPs is to the company CP.
    (2) **Reports:** List reports not covered in the SOPs.
  c. **Signal:** Describe the concept of signal support including current SOI edition or refer to the higher HQ OPORD. Include the following information.
    (1) Identify the **SOI index** in effect.
    (2) Identify **methods of communication by priority**.
    (3) Describe **pyrotechnics and signals** such as hand and arm and demonstrate as appropriate.
    (4) Give **code words** such as OPSKEDs.
    (5) Give the **challenge and password** for use behind friendly lines.
    (6) Give the **number combination** for use forward of friendly lines.
    (7) Give the **running password**.
    (8) Give **recognition signals** (near—far, day—night).

Table 2-11. Example of a fragmentary order format (continued)

Field FRAGORD guidance:
1. The field FRAGORD issuance lasts a target of 30 minutes but not longer than 40 minutes. The proposed planning guide is as follows.
  a. Paras 1 and 2 — 5 minutes.
  b. Para 3 — 20 to 30 minutes.
  c. Paras 4 and 5 — 5 minutes.
2. The FRAGORD focuses on actions on the objective. The platoon leader may use subordinates to prepare paras 1, 4, and 5 and routes and fires for the FRAGORD. Subordinates may brief the portions of the FRAGORD they prepare.
3. The use of sketches and a terrain model is critical to rapid understanding of the operation and FRAGORD.
4. Rehearsals are critical as elements of the constrained planning model. Effective rehearsals coinciding with the FRAGORD reduces preparation time, affording the platoon leader more time for movement and reconnaissance.
5. Planning in a field environment reduces the amount of time leaders have for in-depth mission planning. The troop leading procedures give leaders a framework to plan missions and produce orders when time is short.

Legend: BMNT–begin morning nautical twilight; CP–command post; EENT–end of evening nautical twilight; FRAGORD–fragmentary order; HQ–headquarters; OPORD–operation order; OPSKED–operation schedule; para–paragraph; SOI–signal operating instructions; SOP–standard operating procedure; Temp–temperature

ANNEXES

2-23. OPORD annexes follow the issuance of an OPORD only when more information about truck movement, air assault, PBs, small boats, linkups, or stream crossings becomes necessary. Brevity is standard. (See tables 2-12 through 2-16 on pages 2-26 through 2-35 for sample formats of various types of annexes.)

Table 2-12. Example of an air movement annex format

AIR MOVEMENT ANNEX.
1. **SITUATION.**
  a. **Enemy:** Include the following information.
    (1) Enemy air capability.
    (2) Enemy air defense artillery capability.
    (3) Weather including the percent illumination, illumination angle, night-vision device window, ceiling, and visibility.
 	
2. **MISSION.** Who, what (task), when, where, and why (purpose).
 	
3. **EXECUTION.** Include the following information.
  a. **Concept of operations.**
  b. **Tasks to subordinate units.**

Table 2-12. Example of an air movement annex format (continued)

c. **Coordinating instructions:** Include the following information.
  (1) **PZ.**
    (a) Name and number.
    (b) Coordinates.
    (c) Load time.
    (d) Takeoff time.
    (e) Markings.
    (f) Control.
    (g) Landing formation.
    (h) Approach and departure directions.
    (i) Alternate PZ name and number.
    (j) Penetration points.
    (k) Extraction points.
  (2) **LZ.**
    (a) Name and number.
    (b) Coordinates.
    (c) H-hour.
    (d) Markings.
    (e) Control.
    (f) Landing formation and direction.
    (g) Alternate LZ name and number.
    (h) Deception plan.
    (i) Extraction LZ.
  (3) **Laager site.**
    (a) Communications.
    (b) Security force.
    (c) Flight routes and alternates.
    (d) Abort criteria.
    (e) Downed aircraft/crew designated area of recovery.
    (f) Special instructions.
    (g) Cross-FLOT considerations.
    (h) Aircraft speed.
    (i) Aircraft altitude.
    (j) Aircraft crank time.
    (k) Rehearsal schedule and plan.
    (l) Actions on enemy contact en route and on the ground.
4. **SUSTAINMENT.**
  a. **Logistics:** Include the following information.
    (1) **Sustainment overlay:** Include forward area refuel and rearm points.
    (2) **Maintenance:** This is specific to aircraft.
    (3) **Transportation.**

Table 2-12. Example of an air movement annex format (continued)

4. **SUSTAINMENT** (continued).
    (4) **Supply.**
        (a) Class I—food, rations, and water.
        (b) Class III—petroleum, oils, and lubricants.
        (c) Class V—ammunition and explosives.
        (d) Class VII—major end items.
        (e) Class VIII—medical materiel.
        (f) Class IX—repair parts.
        (g) Distribution methods.

5. **COMMAND AND SIGNAL.**
    a. **Command:** Include the following information.
      (1) **Location of commander or patrol leader:** State where the commander intends to be during the operation, by phase whenever the operation is phased.
      (2) **Succession of command:** State the succession of command when the unit's SOP does not.
    b. **Control:** List reports not covered in the SOPs.
    c. **Signal:** Describe the concept of signal support including current signal operating instructions edition or refer to the higher headquarters' operation order.
      (1) Air and ground call signs and frequencies.
      (2) Air and ground emergency codes.
      (3) Passwords and number combinations.
      (4) Fire net and quick-fire net.
      (5) Time zone.
      (6) Time check.

Legend: FLOT—forward line of own troops; LZ–landing zone; PZ–pickup zone; SOP–standard operating procedure

Table 2-13. Example of a patrol base annex format

PATROL BASE ANNEX.
1. **SITUATION.** Include the following information.
    a. **Enemy forces.**
    b. **Friendly forces.**
    c. **Attachments and detachments.**

2. **MISSION.** Who, what (task), when, where, and why (purpose).

3. **EXECUTION.** Include the following information.
    a. **Concept of operations.**
    b. **Scheme of movement and maneuver.**
    c. **Scheme of fires.**

Table 2-13. Example of a patrol base annex format (continued)

3. **EXECUTION** (continued).
  d. **Tasks to subordinate units.** Include the following information.
    (1) Teams.
      (a) Security.
      (b) Reconnaissance.
      (c) Surveillance.
      (d) Listening and observation posts.
    (2) Individuals.
  e. **Coordinating instructions.** Include the following information.
    (1) Occupation plan.
    (2) Operations plan.
      (a) Security plan.
      (b) Alert plan.
      (c) Priority of work.
      (d) Evacuation plan.
      (e) Alternate patrol base when primary base is unsuitable or compromised.

4. **SUSTAINMENT.** Brief only those specifics not covered in the base order.
  a. **Logistics:** Include the following information.
    (1) **Sustainment overlay:** Include water, maintenance, hygiene, rations, and rest plans.
    (2) **Maintenance.**
    (3) **Transportation.**
    (4) **Supply.**
      (a) Class I—food, rations, and water.
      (b) Class III—petroleum, oils, and lubricants.
      (c) Class V—ammunition and explosives.
      (d) Class VII—major end items.
      (e) Class VIII—medical materiel.
      (f) Class IX—repair parts.
      (g) Distribution methods.
    (5) **Field services.**
  b. **Personnel services support:** Include the following information.
    (1) Methods of marking and handling enemy prisoners of war.
    (2) Religious services.
  c. **Army Health System support:** Include the following information.
    (1) Medical mission command.
    (2) Medical treatment.
    (3) Medical evacuation.
    (4) Preventive medicine.

Table 2-13. Example of a patrol base annex format (continued)

5. **COMMAND AND SIGNAL.**
   a. **Command:** Include the following information.
      (1) **Location of commander or patrol leader:** State where the commander intends to be during the operation, by phase whenever the operation is phased.
      (2) **Succession of command:** State the succession of command when the unit's SOP does not.
   b. **Control:** Include the following information.
      (1) **CPs:** Describe the employment of any CPs including the location of each CP and its times of opening and closing. Typically, at platoon level, any reference to CPs is to the company CP.
      (2) **Reports:** List reports not covered in the SOPs.
   c. **Signal:** Describe the concept of signal support including current SOI edition or refer to the higher headquarters' operation order.
      (1) Identify the **SOI index** in effect.
      (2) Identify **methods of communication by priority**.
      (3) Describe **pyrotechnics and signals** such as hand and arm and demonstrate as appropriate.
      (4) Give **code words** such as OPSKEDs.
      (5) Give the **challenge and password** for use behind friendly lines.
      (6) Give the **number combination** for use forward of friendly lines.
      (7) Give the **running password**.
      (8) Give **recognition signals** (near—far, day—night).

Legend: CP—command post; OPSKED—operation schedule; SOI—signal operating instructions; SOP—standard operating procedure

Table 2-14. Example of a waterborne insertion annex format

WATERBORNE INSERTION ANNEX.
1. **SITUATION.** Include the following information.
   a. **Area of operations.**
      (1) **Terrain.**
         (a) River's width.
         (b) River's depth and water's temperature.
         (c) River's current.
         (d) Vegetation.
      (2) **Weather.**
         (a) Tide.
         (b) Surf.
         (c) Wind.
   b. **Enemy forces:** State any changes or additions to identification, location, activity, and strength.
   c. **Friendly forces** (the unit furnishing support).
   d. **Attachments and detachments.**
   e. **Organization for movement.**
2. **MISSION.** Who, what (task), when, where, and why (purpose).

Table 2-14. Example of a waterborne insertion annex format (continued)

3. **EXECUTION.** Include the following information.
  a. **Concept of operations.**
  b. **Scheme of movement and maneuver.**
  c. **Scheme of fires.**
  d. **Tasks to subordinate units.**
    (1) Security.
    (2) Tie down teams.
      (a) Load equipment.
      (b) Secure equipment.
    (3) Designation of coxswains and boat commanders.
      (a) Selection of navigator(s) and observer(s).
  e. **Coordinating instructions.**
    (1) Formations and order of movement.
    (2) Route and alternate route.
    (3) Method of navigation.
    (4) Actions on enemy contact.
    (5) Rally points.
    (6) Embarkation plan.
    (7) Debarkation plan.
    (8) Rehearsals.
    (9) Time schedule.

4. **SUSTAINMENT.** Brief only those specifics not covered in the base order.
  a. **Logistics:** Include the following information.
    (1) **Sustainment overlay.**
    (2) **Maintenance.**
    (3) **Transportation:** Include the disposition of boats, paddles, and life jackets upon debarkation.
    (4) **Supply.**
      (a) Class I—food, rations, and water.
      (b) Class III—petroleum, oils, and lubricants.
      (c) Class V—ammunition and explosives.
      (d) Class VII—major end items.
      (e) Class VIII—medical materiel.
      (f) Class IX—repair parts.
      (g) Distribution methods: Include the method of distributing paddles and life jackets.

5. **COMMAND AND SIGNAL.**
  a. **Command:** Include the following information.
    (1) **Location of commander or patrol leader:** State where the commander intends to be during the operation, by phase whenever the operation is phased.
    (2) **Succession of command:** State the succession of command when the unit's SOP does not.

Table 2-14. Example of a waterborne insertion annex format (continued)

5. **COMMAND AND SIGNAL** (continued).
    b. **Control:** Include the following information.
        (1) **CPs:** Describe the employment of any CPs including the location of each CP and its times of
            opening and closing. Typically, at platoon level, any reference to CPs is to the company CP.
        (2) **Reports:** List reports not covered in the SOPs.
    c. **Signal:** Describe the concept of signal support including current SOI edition or refer to the higher
    headquarters' operation order.
        (1) Identify the **SOI index** in effect.
        (2) Identify **methods of communication by priority**.
        (3) Describe **pyrotechnics and signals** such as hand and arm and demonstrate as appropriate.
        (4) Give **code words** such as OPSKEDs.
        (5) Give the **challenge and password** for use behind friendly lines.
        (6) Give the **number combination** for use forward of friendly lines.
        (7) Give the **running password**.
        (8) Give **recognition signals** (near—far, day—night).

Legend: CP–command post; OPSKED–operation schedule; SOI–signal operating instructions; SOP–standard operating procedure

Table 2-15. Example of a covert gap crossing annex format

COVERT GAP CROSSING ANNEX.
1. **SITUATION.** Include the following information.
    a. **Area of operations.**
        (1) **Terrain.**
            (a) River's width.
            (b) River's depth and water's temperature.
            (c) River's current.
            (d) Vegetation.
            (e) Obstacles.
        (2) **Weather.**
    b. **Enemy forces:** Include location, identification, and activity.
    c. **Friendly forces.**
    d. **Attachments and detachments.**

2. **MISSION.** Who, what (task), when, where, and why (purpose).

3. **EXECUTION.** Include the following information.
    a. **Concept of operations.**
    b. **Scheme of movement and maneuver.**
    c. **Scheme of fires.**

Table 2-15. Example of a covert gap crossing annex format (continued)

3. **EXECUTION** (continued).
    d. **Tasks to subordinate units.**
        (1) Elements.
        (2) Teams.
        (3) Individuals.
    e. **Coordinating instructions.**
        (1) Crossing procedure and techniques.
        (2) Security.
        (3) Order of crossing.
        (4) Actions on enemy contact.
        (5) Alternate plan.
        (6) Rallying points.
        (7) Rehearsal plan.
        (8) Time schedule.

4. **SUSTAINMENT:** Brief only those specifics not covered in the base order.
    a. **Logistics:** Include the following information.
        (1) **Sustainment overlay.**
        (2) **Maintenance.**
        (3) **Transportation.**
        (4) **Supply.**
            (a) Class I—food, rations, and water.
            (b) Class III—petroleum, oils, and lubricants.
            (c) Class V—ammunition and explosives.
            (d) Class VII—major end items.
            (e) Class VIII—medical materiel.
            (f) Class IX—repair parts.
            (g) Distribution methods.

5. **COMMAND AND SIGNAL.**
    a. **Command:** Include the following information.
        (1) **Location of commander or patrol leader:** State where the commander intends to be during the operation, by phase whenever the operation is phased.
        (2) **Succession of command:** State the succession of command when the unit's SOP does not.
    b. **Control:** Include the following information.
        (1) **CPs:** Describe the employment of any CPs including the location of each CP and its times of opening and closing. Typically, at platoon level, any reference to CPs is to the company CP.
        (2) **Reports:** List reports not covered in the SOPs.

Table 2-15. Example of a covert gap crossing annex format (continued)

5. **COMMAND AND SIGNAL** (continued).
   c. **Signal.** Describe the concept of signal support including current SOI edition or refer to the higher headquarters' operation order.
      (1) Identify the **SOI index** in effect.
      (2) Identify **methods of communication by priority**.
      (3) Describe **pyrotechnics and signals** such as hand and arm and demonstrate as appropriate.
      (4) Give **code words** such as OPSKEDs.
      (5) Give the **challenge and password** for use behind friendly lines.
      (6) Give the **number combination** for use forward of friendly lines.
      (7) Give the **running password**.
      (8) Give **recognition signals** (near—far, day—night).

Legend: CP—command post; OPSKED—operation schedule; SOI—signal operating instructions; SOP—standard operating procedure

Table 2-16. Example of a truck annex format

TRUCK ANNEX.
1. **SITUATION.** Include the following information.
   a. **Enemy situation.**
   b. **Friendly situation.**
   c. **Attachments and detachments.**

2. **MISSION.** Who, what (task), when, where, and why (purpose).

3. **EXECUTION.** Include the following information.
   a. **Concept of operations.**
   b. **Scheme of movement and maneuver.**
   c. **Scheme of fires.**
   d. **Tasks to subordinate units.**
   e. **Coordinating instructions.**
      (1) Times of departure and return.
      (2) Loading plan and order of movement.
      (3) Route (primary and alternate).
      (4) Air guards.
      (5) Actions on enemy contact (vehicle ambush) during movement, loading, and unloading.
      (6) Actions at the de-trucking point.
      (7) Rehearsals.
      (8) Vehicle speed, separation, and recovery plan.
      (9) Broken vehicle instructions.

4. **SUSTAINMENT.** Brief only those specifics not covered in the base order.
   a. **Logistics:** Include the following information.
      (1) **Sustainment overlay.**

Table 2-16. Example of a truck annex format (continued)

4. SUSTAINMENT (continued).
   (2) **Maintenance.**
   (3) **Transportation.**
   (4) **Supply.**
       (a) Class I—food, rations, and water.
       (b) Class III—petroleum, oils, and lubricants.
       (c) Class V—ammunition and explosives.
       (d) Class VII—major end items.
       (e) Class VIII—medical materiel.
       (f) Class IX—-repair parts.
       (g) Distribution methods.
   (5) **Field services.**
   b. **Personnel services support:** Include the following information.
   (1) Methods of marking and handling enemy prisoners of war.
   (2) Religious services.
   c. **Army Health System support.** Include the following information.
   (1) Medical mission command.
   (2) Medical treatment.
   (3) Medical evacuation.
   (4) Preventive medicine.

5. COMMAND AND SIGNAL.
   a. **Command:** Include the following information.
   (1) **Location of commander or patrol leader:** State where the commander intends to be during the operation, by phase whenever the operation is phased.
   (2) **Succession of command:** State the succession of command when the unit's SOP does not.
   b. **Control:** Include the following information.
   (1) **CPs:** Describe the employment of any CPs including the location of each CP and its times of opening and closing. Typically, at platoon level, any reference to CPs is to the company CP.
   (2) **Reports:** List reports not covered in the SOPs.
   c. **Signal:** Describe the concept of signal support including current SOI edition or refer to the higher headquarters' operation order. Include the following information.
   (1) Identify the **SOI index** in effect.
   (2) Identify **methods of communication by priority**.
   (3) Describe **pyrotechnics and signals** such as hand and arm and demonstrate as appropriate.
   (4) Give **code words** such as OPSKEDs.
   (5) Give the **challenge and password** for use behind friendly lines.
   (6) Give the **number combination** for use forward of friendly lines.
   (7) Give the **running password**.
   (8) Give **recognition signals** (near—far, day—night).

Legend: CP—command post; OPSKED—operation schedule; SOI—signal operating instructions; SOP—standard operating procedure

COORDINATION CHECKLISTS

2-24. A PL or an SL checks specific items when planning for a combat operation. (See tables 2-17 through 2-24 on pages 2-36 through 2-42 for sample checklists of these items.) In some cases, the PL or SL coordinates directly with the appropriate staff section. In most cases, the company commander or PL provides this information. The PL or SL sometimes carries copies of these checklists to keep from overlooking any vital element to the mission.

Table 2-17. Intelligence coordination checklist

INTELLIGENCE COORDINATION CHECKLIST.
The unit one level up constantly updates intelligence. This ensures the platoon leader's plan reflects the most recent enemy activity. Include the following information in the coordination brief.
1. Identification of enemy unit.
2. Weather and light data.
3. Terrain update.
a. Aerial photos.
b. Trails and obstacles not appearing on the map.
4. Known or suspected enemy locations.
5. Weapons.
6. Probable course of action.
7. Most dangerous course of action.
8. Recent enemy activities.
9. Reaction time of reaction forces.
10. Civilians in the battlespace.
11. Updates to the commander's critical information requirement(s).

Table 2-18. Operations coordination checklist

OPERATIONS COORDINATION CHECKLIST.
The platoon or squad leader coordinates with the company commander or platoon leader to confirm the mission and operations plan, receive last-minute changes, and update subordinates in person or by issuing a fragmentary order. Include the following information in the coordination brief.
1. Mission confirmation brief.
2. Identification of friendly units.
3. Changes in the friendly situation.
4. Route selection and landing, pickup, and drop zone selections.
5. Commander's critical information requirement(s).
a. Changes to priority intelligence requirement(s).
b. Changes to friendly force information requirement(s).
c. Changes to essential elements of friendly information.
d. Changes to rules of engagement.

Table 2-18. Operations coordination checklist (continued)

6. Linkup procedures.
a. Contingencies.
b. QRF.
c. QRF frequency.
7. Transportation and movement plan.
8. Resupply (with S–4).
9. Signal plan.
10. Departure and reentry of forward units.
11. Special equipment requirements.
12. Adjacent units in the area of operations.
13. Rehearsal areas.
14. Methods of insertion and extraction.
Legend: QRF–quick reaction force; S–4–battalion or brigade logistics staff officer

Table 2-19. Fire support coordination

FIRE SUPPORT COORDINATION CHECKLIST.
The platoon or squad leader coordinates the following with the forward observer at squad level and the fire support officer at platoon level. Include the following information in the coordination brief.
1. Mission confirmation brief.
2. Identification of the supporting unit.
3. Mission and objective.
4. Routes including alternate routes to and from the objective.
5. Time of departure and expected time of return.
6. Unit target list from the fire plan.
7. Types of available support (for example, artillery, mortar, naval gunfire, aerial support—including Army, Navy, and Air Force) and their locations
8. Ammunition available including different fuzes.
9. Priority of fires.
10. Control measures.
   a. Checkpoints.
   b. Boundaries.
   c. Phase lines.
   d. Fire support coordination measures.
   e. Priority targets (target list).
   f. Restrictive fire area.
   g. Restrictive fire line.
   h. No-fire area.
   i. Precoordinated authentication.
11. Communication including primary and alternate means, emergency signals, and code words.

Table 2-20. Coordination with forward unit checklist

COORDINATION WITH FORWARD UNIT CHECKLIST.
A platoon or squad requiring foot movement through a friendly forward unit coordinates with that unit's commander for a safe, orderly passage. When no designated time and place for coordination with the forward unit exists, the platoon or squad leader sets a time and place to coordinate with the S–3. The leader speaks to a forward unit member with the authority to commit that forward unit to assisting the platoon or squad during departure. Coordination is a two-way exchange of information. Include the following information in the coordination brief.
1. Identification of yourself and your unit.
2. Size of the platoon or squad.
3. General area of operations.
4. Known or suspected enemy positions or obstacles.
5. Possible enemy ambush sites.
6. Latest enemy activity.
7. Detailed information on friendly positions such as crew-served weapons and final protective fire.
8. Fire and barrier plan.
   a. Support the unit is capable of furnishing such as what they do and for how long.
      (1) Fire support.
      (2) Aid and litter teams.
      (3) Navigational signals and aids.
      (4) Guides.
      (5) Communications.
      (6) Reaction units.
      (7) Other.
   b. Call signs and frequencies.
   c. Pyrotechnic plan.
   d. Challenge and password, running password, and number combination.
   e. Emergency signals and code words.
   f. Instruction to pass information to any relieving unit.
   g. Recognition signals.

Legend: S–3–battalion or brigade operations staff officer

Table 2-21. Adjacent unit coordination checklist

ADJACENT UNIT COORDINATION CHECKLIST.
Immediately after the operation order or mission briefing, the platoon or squad leader checks with other platoon and squad leaders who will be operating in the same areas. When the leader is unaware of any other units operating in the area, the leader checks with the S–3 during the operations coordination. The S–3 can help arrange any necessary coordination. The platoon and squad leaders exchange the following information with other units operating in the same area.
1. Identification of the unit.
2. Mission and size of the unit.
3. Planned times and points of departure and reentry.

Table 2-21. Adjacent unit coordination checklist (continued)

4. Route(s).
5. Fire support and control measures.
6. Frequencies and call signs.
7. Challenge and password, running password, and number combination.
8. Pyrotechnic plan.
9. Any information the unit has about the enemy.
10. Recognition signals.
Legend: S–3–battalion or brigade operations staff officer

Table 2-22. Rehearsal area coordination checklist

REHEARSAL AREA COORDINATION CHECKLIST.
The assistant patrol leader coordinates the use of the rehearsal area to facilitate the unit's safe, efficient, and effective use of the area before its mission. Include the following information in the coordination brief.
1. Identification of the unit.
2. Mission.
3. Terrain similar to that of the objective site.
4. Security of the area.
5. Availability of aggressors.
6. Use of blanks, pyrotechnics, and ammunition.
7. Mockups available.
8. Time the area is available, preferably when the light conditions approximate those of the patrol.
9. Transportation.
10. Coordination with other units using the area.

Table 2-23. Army aviation coordination checklist

ARMY AVIATION COORDINATION CHECKLIST.
The patrol leader coordinates this with the company commander or S–3 Air to facilitate the time and detailed and effective use of aviation assets as they apply to the tactical mission:
1. **SITUATION.** Include the following information.
a. **Enemy.**
(1) Air capability.
(2) ADA capability.
(3) Weather including the percent of illumination, illumination angle, night-vision device window, ceiling, and visibility.
b. **Friendly.**
(1) Unit(s) supporting the operation and axis of movement/corridor/routes.
(2) ADA status.
2. **MISSION.** Who, what (task), when, where, and why (purpose).

Table 2-23. Army aviation coordination checklist (continued)

3. **EXECUTION.** Include the following information.
  a. **Concept of operations:** Overview of what the requesting unit wants to accomplish with the air assault / air movement.
  b. **Tasks to combat units.**
    (1) Infantry.
    (2) Attack aviation.
  c. **Tasks to combat support units.**
    (1) Artillery.
    (2) Aviation (lift).
  d. **Coordinating instructions.**
    (1) PZ.
      (a) Direction of landing.
      (b) Time of landing and flight direction.
      (c) Locations of PZ and alternate PZ.
      (d) Loading procedures.
      (e) Marking of PZ (for example, by panel, smoke, signal mirror, lights).
      (f) Planned flight route including start point, air control point, and release point.
      (g) Formations (PZ, en route, LZ).
      (h) Code words.
        _1_ PZ secure (before landing), PZ clear (lead bird and last bird).
        _2_ Alternate PZ (PZ, en route. LZ). names of PZ and alternate PZ.
      (i) TAC air and artillery.
      (j) Number of passengers for each helicopter and for the entire lift.
      (k) Equipment carried by individuals.
      (l) Marking of key leaders.
      (m) Abort criteria (PZ, en route, LZ).
    (2) LZ.
      (a) Direction of landing.
      (b) False insertion plans.
      (c) Time of landing in LZ's time zone.
      (d) Locations of LZ and alternate LZ.
      (e) Marking of LZ (for example, by panel, smoke, signal mirror, lights).
      (f) Formation of landing.
      (g) Code words and names of LZ and alternate LZ.
      (h) TAC air and artillery preparation. fire support coordination.
        _1_ Secure LZ?
        _2_ Do not secure LZ?

Table 2-23. Army aviation coordination checklist (continued)

4. **SUSTAINMENT.** Brief only those specifics not covered in the base order. Include the number of aircraft for each lift, number of lifts, whether the aircraft will refuel or rearm during the mission, special equipment carried by personnel, aircraft configuration, a bump plan, and the following info.
  a. **Logistics.**
    (1) **Sustainment overlay.**
    (2) **Maintenance.**
    (3) **Transportation.**
    (4) **Supply.**
      (a) Class I—food, rations, and water.
      (b) Class III—petroleum, oils, and lubricants.
      (c) Class V—ammunition and explosives.
      (d) Class VII—major end items.
      (e) Class VIII—medical materiel.
      (f) Class IX—repair parts.
      (g) Distribution methods.
    (5) **Field services.**
  b. **Personnel services support.**
    (1) Methods of marking and handling enemy prisoners of war.
    (2) Religious services.
  c. **Army Health System support.**
    (1) Medical mission command.
    (2) Medical treatment.
    (3) Medical evacuation.
    (4) Preventive medicine.
5. COMMAND AND SIGNAL.
  a. **Command:** Include the following information.
    (1) **Location of commander or patrol leader:** State where the commander intends to be during the operation, by phase whenever the operation is phased. Also include the locations of the air mission commander, ground tactical commander, and air assault task force commander.
    (2) **Succession of command:** State the succession of command when the unit's SOP does not.
  b. **Control.** Include the following information.
    (1) **CPs:** Describe the employment of any CPs including the location of each CP and its times of opening and closing. Typically, at platoon level, any reference to CPs is to the company CP.
    (2) **Reports:** List reports not covered in the SOPs.

Table 2-23. Army aviation coordination checklist (continued)

5. **COMMAND AND SIGNAL** (continued).
c. **Signal:** Describe the concept of signal support including current SOI edition or refer to the higher headquarters' operation order.
(1) Identify the **SOI index** in effect.
(2) Identify **methods of communication by priority**.
(3) Describe **pyrotechnics and signals** such as hand and arm and demonstrate as appropriate.
(4) Give **code words** such as OPSKEDs.
(5) Give the **challenge and password** for use behind friendly lines.
(6) Give the **number combination** for use forward of friendly lines.
(7) Give the **running password**.
(8) Give **recognition signals** (near—far, day—night).
Legend: ADA–air defense artillery; CP–command post; LZ–landing zone; OPSKED–operation schedule; PZ–pickup zone; S–3–battalion or brigade operations staff officer; SOI–signal operating instructions; SOP–standard operating procedure; TAC–tactical

Table 2-24. Vehicular movement coordination checklist

VEHICULAR MOVEMENT COORDINATION CHECKLIST.
The platoon or first sergeant coordinates this with the supporting unit to facilitate the effective, detailed, and efficient use of vehicular support and assets. Include the following information in the coordination brief.
1. Identification of the unit.
2. Identification of the supporting unit.
3. Number and type of vehicles and tactical preparation.
4. Entrucking point.
5. Departure time.
6. Preparation of vehicles for movement.
a. Driver's responsibilities.
b. Platoon's and squad's responsibilities.
c. Requisite special supplies and equipment.
7. Availability of vehicles for preparation, rehearsals, and inspection, as well as times and locations.
8. Routes.
a. Primary.
b. Alternate.
c. Checkpoints.
9. De-trucking points.
a. Primary.
b. Alternate.
10. Order of march.
11. Speed.
12. Communications including frequencies, call signs, and codes.
13. Emergency procedures and signals.

ACTIONS, PURPOSE, OPERATIONS, AND TASKS

2-25. Friendly forces take actions with the intention of producing the effects those actions are designed to have on the enemy. (See table 2-25 for those actions.) Purpose is the desired or intended result of the tactical operation stated in terms related to the enemy or the desired situation. Purpose is the _why_ of the mission statement and often follows the words _in order to_. It is the most important component of the mission statement. (See table 2-26.)

Table 2-25. Tactical mission tasks

Attack by fire	Control	Follow and assume	Retain
Block	Counterreconnaissance	Follow and support	Secure
Breach	Destroy	Interdict	Seize
Bypass	Disengage	Isolate	Support by fire
Canalize	Disrupt	Neutralize	Suppress
Clear	Exfiltrate	Occupy	Turn
Contain	Fix	Reduce	

Table 2-26. Purpose of mission statement

Allow	Divert	Prevent
Cause	Enable	Protect
Create	Envelop	Support
Deceive	Influence	Surprise
Deny	Open	

2-26. An operation is a military action or the carrying out of a military action to gain the objectives of any given battle or campaign. Rangers accomplish operations through tasks. A task is a specific, clearly defined, decisive, and measurable activity or action accomplished by a Ranger or organization that contributes to the achievement of encompassing missions or other requirements. (See table 2-27 on page 2-44.) Also, some shaping tasks are vital to the security of Soldiers and the success of missions. (See table 2-28 on page 2-44.)

Table 2-27. Elements of offense and of defense

ELEMENTS OF OFFENSE AND SUBORDINATE TASKS	
Offensive Tasks	**Defensive Tasks**

Variations of the movement to contact:
   Search and attack
   Cordon and search
	Area defense
	
	Variations of the area defense:
	   Defense of a linear obstacle
	   Perimeter defense
	   Reverse slope defense

Attack
 	
Variations of the attack:
   Ambush
   Counterattack
   Raid
   Spoiling attack
	Mobile defense

ELEMENTS OF DEFENSE AND SUBORDINATE TASKS	
Offensive Tasks	**Defensive Tasks**

Movement to contact
Attack
Exploit
Pursuit
 	
Forms of maneuver:
   Frontal attack
   Penetration
   Envelopment
   Turning movement
   Infiltration
	Retrograde
	
	Variations of the retrograde:
	   Delay
	   Withdrawal
	   Retirement

Table 2-28. Enabling operations

Reconnaissance:
   Zone
   Area
   Route
   Reconnaissance in force
   Special
	Security:
	   Screen
	   Guard
	   Cover
	   Area security
		Troop movement:
		   Nontactical movement
		   Tactical movement

Relief in place	Passage of lines	Countermobility
Mobility	Tactical deception	Linkup

TERRAIN MODEL

2-27. During the planning process, the terrain model (see figure 2-2 on page 2-46) offers an effective visual communication of the patrol routes and detail actions on the objective. At a minimum, the model is useful for displaying routes to the objective and highlighting prominent terrain features the patrol encounters during movement. A second terrain model of the objective area also supports the planning process. It is large and detailed enough to brief the patrol's actions on the objective.

    a. <u>Checklist</u>. Include the following items in the terrain models.

        (1) North-seeking arrow.

        (2) Scale.

        (3) Grid lines.

        (4) Objective location.

        (5) Exaggerated terrain relief and water obstacles.

        (6) Friendly patrol locations.

        (7) Targets (indirect fire including grid and type of round).

        (8) Primary and alternate routes.

        (9) Planned release points (RPs): objective rally point, linkup rally point, RP.

        (10) Danger areas such roads, trails, and open areas.

        (11) Legend.

        (12) Blowup of objective area.

    b. <u>Construction</u>. Some field-expedient techniques for constructing terrain models follow.

        (1) Use a 3-inch by 5-inch card from a ready-to-eat meal box or a piece of paper to label the objective or key sites.

        (2) Use colored tape or string from the guts of parachute cord to make grid lines. Identify the grids with numbers written on small pieces of paper.

        (3) Replicate trees and vegetation using moss, green or brown spray paint, pine needles, crushed leaves, or cut grass.

        (4) Use blue chalk, blue spray paint, blue yarn, tin foil, or creamer from a ready-to-eat meal to designate bodies of water.

        (5) Make north-seeking arrows from sharpened twigs, pencils, or colored yarn.

        (6) Use red yarn, 5.56-millimeter (mm) rounds, toy soldiers, or poker chips to designate enemy positions.

        (7) Construct friendly positions such as security elements, support by fire, and assault elements using 5.56-mm rounds, toy soldiers, poker chips, sugar or coffee packets from ready-to-eat meals, or preprinted acetate cards.

        (8) Use small pieces of cardboard or paper to identify target reference points and indirect fire targets. Show the grids for each point.

        (9) Construct breach, support by fire, and assault positions using the same methods and again using colored yarn or string for easy identification.

        (10) Construct bunkers and buildings using ready-to-eat meal boxes, tongue depressors, or sticks.

        (11) Construct perimeter wire from a spiral notebook.

        (12) Construct key phase lines with colored string or yarn.

        (13) Use colored tape or yarn to replicate trench lines by digging a furrow and tinting it with colored chalk or spray paint.

Note. Clearly identify in a legend all the symbols on the terrain model.

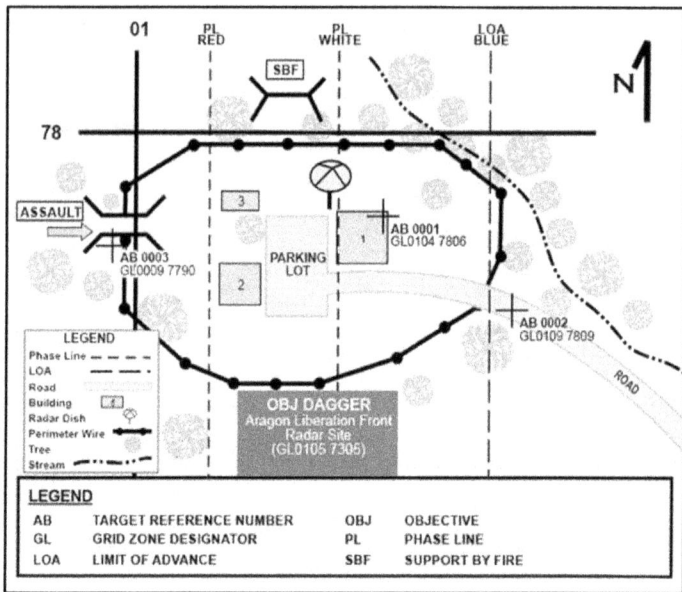

Figure 2-2. Terrain model

Chapter 3

Fire Support

Indirect fire support greatly increases the combat effectiveness and survivability of any Infantry unit. Every Ranger and small-unit leader masters planning for and using this asset effectively. Fire support assets help a unit by suppressing, fixing, destroying, or neutralizing the enemy. Leaders consider employing indirect fire support throughout every offensive and defensive operation. This chapter discusses plans, tasks, capabilities, risk estimate distances (REDs), target overlays, close air support (CAS) elements, and sequences of calls for fire. It also provides an example of call for fire transmissions. (See appendix A for more information on fire support.)

BASIC TASKS AND TARGETING

3-1. The effectiveness of the fire support system depends on successful performance of its four basic tasks. These include supporting forces in contact, supporting the battle plan, synchronizing the fire support system, and sustaining the fire support system. Targeting objectives is the overall effect the leader hopes to achieve with fire support assets. The first functional step in the targeting process is to decide. A decision defines the overall focus and sets priorities for collecting intelligence and planning the attack. The leader addresses targeting priorities for each phase or critical event of an operation. At all echelons, one or more alternative COA(s) is/are analyzed. Each is based on mission analysis, current and projected battle situations, and anticipated opportunities.

3-2. The second critical function is to detect. The intelligence officer (either the assistant chief of staff for intelligence or the battalion or brigade intelligence staff officer) directs the effort to detect the high-payoff targets identified during the step to decide. To identify the exact who, what, when, and how of target acquisition, the intelligence officer works closely with the analysis and control element, field artillery intelligence officer, targeting officer, and fire support officer.

INTERDICTION

3-3. Interdiction is an action to limit, divert, disrupt, delay, destroy, and damage the enemy's military surface capabilities such as vessels, vehicles, aircraft, personnel, and cargo before their effective use against friendly forces succeeds, or to otherwise achieve friendly objectives. Table 3-1 on page 3-2 and table 3-2 on page 3-3 outline the capabilities of mortars and field artillery. Interdiction—

- Limits — Reduces enemy options (for example, direct air interdiction and fire support to limit the enemy's AAs and fire support).
- Diverts — Creates a distraction that forces the enemy to tie up critical resources (for example, an attack on targets that causes the enemy to move capabilities or assets from one area or activity to another).
- Disrupts — Stops effective interaction between the enemy and their support systems, reduces enemy efficiency, and increases their vulnerability.
- Delays — Disrupts, diverts, or destroys enemy capabilities or targets. Changes the enemy's abilities to reach a certain point in the battlespace or to project combat power from it.

- Destroys — Ruins the structure or condition of a vital enemy target. Destruction is definable as an objective by providing a quantity or percentage of an enemy asset or target the friendly weapon system(s) can realistically destroy. For example, artillery normally states destruction comprises a 30-percent reduction in capability or structural integrity; maneuver combat forces normally claim 70 percent.
- Damages — Effects battle damage on the objective, which a subsequent subjective or objective assessment deems light, moderate, or severe.

Table 3-1. Capabilities of mortars

Weapon	Munition Available	Maximum Range	Minimum Range	Maximum Rate of Fire	Sustained Rate of Fire	Burst Diameter
60-mm	HE, WP, Illum	3,490 m (HE)	70 m (HE)	30 RPM for 4 minutes	15–20 RPM (depends on ammo)	30 m
81-mm	HE, WP, Illum	5,608 m (HE)	83 m (HE)	25–30 RPM for 2 minutes (depends on ammo)	8–15 RPM (depends on ammo)	38 m
120-mm	HE, Smoke, Illum	7,200 m (HE)	200 m (HE)	16 RPM for 1 minute	4 RPM	60 m

Note: See TM 9-1010-233-10 for the 60-mm mortar's sustained rates of fire by ammunition type. See TM 9-1015-249-10 and TM 9-1015-257-10 for the 81-mm mortar's maximum and sustained rates of fire by ammunition type.

Legend: ammo–ammunition; HE–high explosive; Illum–illumination; m–meter; mm–millimeter; RPM–rounds per minute; WP–white phosphorous

Table 3-2. Capabilities of field artillery

ARTILLERY	AMMUNITION		RANGE (meters)			RATES OF FIRE (rounds per minute)	
	Projectile	Fuze	Maximum	DPICM	RAP	Sustained	Maximum
105-mm M119	HE, WP, Illum, Illum (IR), DPICM	PD, VT, MT, ET, MTSQ, Delay	11,500	14,100	19,500	3	10
155-mm M198	HE, WP, Illum, Illum (IR), DPICM, M825 smoke, SCATMINE		18,300 or 22,000 with M795 HE, M825 smoke	18,000 M483 or 28,200 with M864	30,100	2	4 for 2 minutes then 2
155-mm M109A5 / A6 / A7	HE, WP, Illum, Illum (IR), DPICM, M825 smoke, SCATMINE	PD, VT, MT, ET, MTSQ, Delay, PGK	18,200 or 21,700 with M795 HE, M825 smoke; 24,500 with M982 Block 1-1a1	17,900 M483 or 28,100 with M864	30,000	1	4 for 2 minutes then 1
155-mm M777-series			22,200 with M201A1 charge 8S or 22,500 with M232, Zone 5; 24,500 with M982 Block 1-1a.	17,900 or 28,100 with M864	30,000	2	4 for 2 minutes then 2

Note: Excalibur rounds are not authorized for the M109A5.

Legend: DPICM—dual-purpose improved conventional munitions; ET—electronic time; HC—hexachloroethane; HE—high explosive; Illum—illumination; IR—infrared; mm—millimeter; MT—mechanical time; MTSQ—mechanical time and super quick; PD—point detonating; PGK—precision guidance kit; RAP—rocket assisted projectile; SCATMINE—scatterable mine; VT—variable time; WP—white phosphorus

DANGER

For mortars and field artillery, when the target is within 600 meters of any friendly troops, announce DANGER CLOSE in the method of engagement portion of the call for fire.

When adjusting 5-inch or smaller naval guns on targets within 750 meters, announce DANGER CLOSE. For larger naval guns on targets within 1,000 meters, announce DANGER CLOSE. Failure to adhere to this guidance can result in fratricide.

Avoid making corrections in using the bracketing method of adjustment. Doing so can result in personnel death or serious injury. Use only the creeping method of adjustment during danger close missions. Make corrections of no more than 100 meters by creeping the rounds to the target.

RISK ESTIMATE DISTANCE

3-4. RED applies to combat only, whereas minimum safe distances apply to training. RED takes into account the bursting radius of particular munitions and the characteristics of the delivery system. It associates this combination with a percentage likelihood of becoming a casualty. This is the percentage of risk.

WARNING

Use RED formulas only in combat to determine acceptable risk levels and to identify the risk to Rangers at various distances from the targets. When training, use minimum safe distances. (See AR 385-63, DA Pam 385-63, and ATP 5-19.)

3-5. RED is defined as the minimum distance friendly troops can approach the effects of friendly fire without suffering appreciable casualties of a 0.1-percent or higher point of impact. Table 3-3 gives the REDs for mortars and cannon artillery and includes the dual-purpose improved conventional munitions. (See ATP 3-09.32 for information on REDs.)

Table 3-3. Risk estimate distances for unguided mortars and cannon artillery

UNGUIDED MORTAR RISK ESTIMATE DISTANCES					
System	Description	Danger Close	Range	0.1 percent PI	
				Standing	Prone
M224 M224A1	60-mm mortar	600 m	1/3 2/3 max	115 m / 378 ft 125 m / 410 ft 145 m / 476 ft	115 m / 378 ft 120 m / 394 ft 145 m / 476 ft
M252 M252A1	81-mm mortar	600 m	1/3 2/3 max	170 m / 558 ft 195 m / 640 ft	160 m / 525 ft 185 m / 607 ft 190 m / 623 ft
M120 M120A1 M121	120-mm mortar	600 m	1/3 2/3 max	280 m / 919 ft 395 m / 1,296 ft 430 m / 1,411 ft	260 m / 853 ft 365 m / 1,198 ft 410 m / 1,345 ft
UNGUIDED CANNON ARTILLERY RISK ESTIMATE DISTANCES					
System	Description	Danger Close	Range	0.1 percent PI	
				Standing	Prone
M119 M119A2	105-mm howitzer HE (M1 Comp B/M760)	600 m	1/3 2/3 max	290 m / 951 ft 300 m / 984 ft 455 m / 1,493 ft	270 m / 886 ft 285 m / 935 ft 430 m / 1,411 ft
	105-mm howitzer HERA (M913 HERA / M297 HERA)	600 m	1/3 2/3 max	250 m / 820 ft 410 m / 1,345 ft 650 m / 2,133 ft	230 m / 755 ft 285 m / 935 ft 430 m / 1,411 ft

Table 3-3. Risk estimate distances for unguided mortars and cannon artillery (continued)

UNGUIDED CANNON ARTILLERY RISK ESTIMATE DISTANCES					
System	Description	Danger Close	Range	0.1 percent PI	
				Standing	Prone
M109A6 M777A2	155-mm howitzer HE (M107 Comp B/M795)	600 m	1/3 2/3 max	300 m / 984 ft 460 m / 1,509 ft 695 m / 2,280 ft	285 m / 935 ft 440 m / 1,444 ft 665 m / 2,182 ft
	155-mm howitzer DPICM (M483A1)	600 m	1/3 2/3 max	270 m / 886 ft 325 m / 1,066 ft 510 m / 1,673 ft	260 m / 853 ft 310 m / 1,017 ft 490 m / 1,608 ft
	155-mm howitzer DPICM (M864)	600 m	1/3 2/3 max	325 m / 1,066 ft 500 m / 1,640 ft 825 m / 2,707 ft	305 m / 1,001 ft 485 m / 1,591 ft 775 m / 2,543 ft
	155-mm howitzer RAP (M945A1 RAP)	600 m	1/3 2/3 max	360 m / 1,181 ft 530 m / 1,739 ft 1,045 m / 3,428 ft	360 m / 1,181 ft 520 m / 1,706 ft 965 m / 3,166 ft

Note: See ATP 3-09.32, appendix I for information on danger close.

Legend: DPICM–dual-purpose improved conventional munitions; ft–feet; HE–high explosive; HERA–high explosive, rocket assisted; m–meter; max–maximum; mm–millimeter; PI–probability of incapacitation; RAP–rocket assisted projectile

3-6. Casualty criterion is the 5-minute assault criterion for a prone Ranger in winter clothing and helmet. Physical incapacitation means a Ranger is physically unable to function in an assault within a 5-minute period after an attack. A point of impact value of less than 0.1 percent is interpretable as less than or equal to one chance in one thousand.

3-7. Echeloning fires within the specified RED for a delivery system requires the unit to assume some risk. The maneuver commander determines by delivery system how close to the forces to allow the fires to fall. Although the decision is made at this risk level, the commander relies heavily on the fire support officer's expertise.

TARGET OVERLAYS

3-8. Table 3-4 lists the contents of fire support overlay. Figure 3-1 depicts the sterile fire support overlay. Figure 3-2 on page 3-8 illustrates the nonsterile fire support overlay, which includes target symbols and index marks to position the overlay on a map.

Table 3-4. Contents of fire support overlay

Unit and official capacity of person making overlay.	Primary and alternate routes.
Date of overlay preparation.	Phase lines and checkpoints used by patrol.
Map sheet number.	Spares.
Effective period of overlay (day, time, group).	Index marks to position overlay on map.
Priority target.	Objective.
Location of objective rally point.	Target symbols.
Call signs and frequencies (primary and alternate).	Completed description, location, and remarks column.

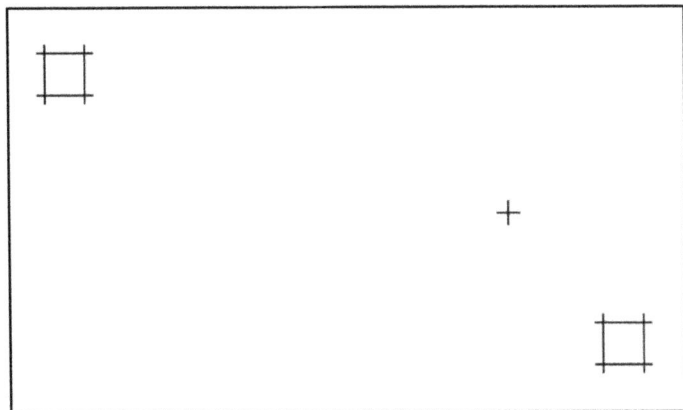

Figure 3-1. Sterile fire support overlay

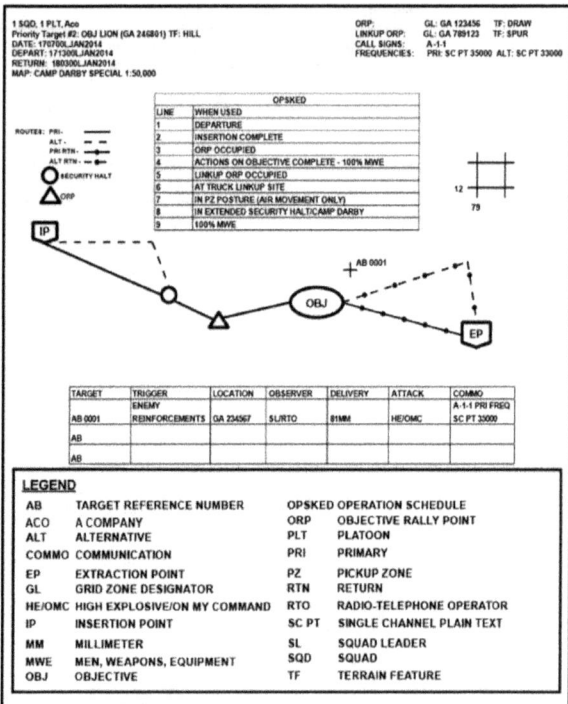

1 SQD, 1 PLT, Aco
Priority Target #2: OBJ LION (GA 246801) TF: HILL
DATE: 170700LJAN2014
DEPART: 171300LJAN2014
RETURN: 180300LJAN2014
MAP: CAMP DARBY SPECIAL 1:50,000

ORP: GL: GA 123456 TF: DRAW
LINKUP ORP: GL: GA 789123 TF: SPUR
CALL SIGNS: A-1-1
FREQUENCIES: PRI: SC PT 35000 ALT: SC PT 33000

ROUTES: PRI -
ALT -
PRI RTN -
ALT RTN -
SECURITY HALT
ORP

OPSKED	
LINE	WHEN USED
1	DEPARTURE
2	INSERTION COMPLETE
3	ORP OCCUPIED
4	ACTIONS ON OBJECTIVE COMPLETE - 100% MWE
5	LINKUP ORP OCCUPIED
6	AT TRUCK LINKUP SITE
7	IN PZ POSTURE (AIR MOVEMENT ONLY)
8	IN EXTENDED SECURITY HALT/CAMP DARBY
9	100% MWE

IP

AB 0001

OBJ

EP

TARGET	TRIGGER	LOCATION	OBSERVER	DELIVERY	ATTACK	COMMO
AB 0001	ENEMY REINFORCEMENTS	GA 234567	SL/RTO	81MM	HE/OMC	A-1-1 PRI FREQ SC PT 35000
AB						
AB						

LEGEND

AB	TARGET REFERENCE NUMBER	OPSKED	OPERATION SCHEDULE
ACO	A COMPANY	ORP	OBJECTIVE RALLY POINT
ALT	ALTERNATIVE	PLT	PLATOON
COMMO	COMMUNICATION	PRI	PRIMARY
EP	EXTRACTION POINT	PZ	PICKUP ZONE
GL	GRID ZONE DESIGNATOR	RTN	RETURN
HE/OMC	HIGH EXPLOSIVE/ON MY COMMAND	RTO	RADIO-TELEPHONE OPERATOR
IP	INSERTION POINT	SC PT	SINGLE CHANNEL PLAIN TEXT
MM	MILLIMETER	SL	SQUAD LEADER
MWE	MEN, WEAPONS, EQUIPMENT	SQD	SQUAD
OBJ	OBJECTIVE	TF	TERRAIN FEATURE

Figure 3-2. Nonsterile fire support overlay

3-9. Using the checklist of target, trigger, location, observer, delivery system, attack guidance, and communications network (see table 3-5) helps ensure the leader's fire support plan is complete. It is useful for identifying all aspects of individual targets before coordination and the OPORD.

Table 3-5. Checklist of target–trigger–location–observer–delivery system–attack guidance– communications network

Target	Number or type of targets
Trigger	When to fire the target
Location	Minimum of six-digit grid
Observer	Primary and alternate
Delivery system	Mortar, artillery, air
Attack guidance	Ammo, special instructions
Communications network	Company TAC, arty CFL
Note: See ATP 3-09.32 for information on fire support.	
Legend: ammo–ammunition; arty CFL–artillery coordinated fire line; TAC–tactical air controller	

CALL FOR FIRE

3-10. A call for fire involves specific steps, starting with call signs and the WARNORD. These steps are precise to preserve both the safety of Rangers and the accurate hits on the target. Asterisks (*) below indicate requisite elements for a basic call for fire mission. The format for delivering a call for fire follows.

    a. Observer's identification. Call signs.*

    b. WARNORD.*

        (1) Type of mission.

            (a) Adjust fire.

            (b) Fire for effect.

            (c) Suppress.

            (d) Immediate suppression / immediate smoke.

        (2) Size of element to fire for effect. When the observer does not specify the size of element to fire, the battalion fire direction center (FDC) decides.

    c. Method of target location.*

        (1) Polar plot.

        (2) Shift from a known point.

        (3) Grid.

        (4) Laser grid.

        (5) Laser plot.

    d. Location of target.*

        (1) Grid coordinates. Requisite six-digit grid and altitude. For greater accuracy, eight- or ten-digit grid.

        (2) Shift from a known point. Send the observer the target direction.

            (a) Milliradians (nearest 10).

            (b) Degrees.

        (c) Cardinal direction.

        (d) Send lateral shift, right/left, nearest 10 meters.

        (e) Send range shift, add/drop, nearest 100 meters.

        (f) Send vertical shift, up/down, nearest 5 meters. Use only when it exceeds 35 meters.

    (3) Polar plot.

        (a) Send direction to nearest 10 milliradians.

        (b) Send distance to nearest 100 meters.

        (c) Send vertical shift to nearest 5 meters.

  e. Description of target *

    (1) Type.

    (2) Activity.

    (3) Number.

    (4) Degree of protection.

    (5) Size and shape (length, width, or radius).

  f. Method of engagement.

    (1) Type of adjustment. When the observer does not request a specific type of fire control adjustment, issue area fire.

        (a) Precision fire = point target.

        (b) Area fire = moving target.

    (2) Danger close. This condition exists when friendly troops are within—

        (a) 600 meters for mortars.

        (b) 600 meters for artillery.

        (c) 750 meters for naval guns, 5 inches or smaller.

    (3) Mark (to orient observer or indicate targets).

    (4) Trajectory.

        (a) Low angle (standard).

        (b) High angle (mortar fire or as requested).

    (5) Ammunition. Use high explosive (HE) projectiles and quick fuzes unless specified by the observer.

        (a) Projectile (HE, illumination, improved conventional munitions, smoke).

        (b) Fuze (quick, timed, or other options).

        (c) Volume of fire. The observer may request the number of rounds to fire.

    (6) Distribution.

        (a) 100-meter sheaf (standard).

        (b) Converged sheaf (small, hard targets).

        (c) Special sheaf (any length, width, and attitude).

        (d) Open sheaf (separate bursts).

        (e) Parallel sheaf (linear target).

  g. Method of fire and of control.

    (1) Method of fire. Specific guns and a specific interval between rounds. Normally adjust fire: one gun is useful with a 5-second interval between rounds.

    (2) Method of control.

        (a) AT MY COMMAND, FIRE. Remains in effect until the observer orders CANCEL AT MY COMMAND.

(b) CANNOT OBSERVE. The observer cannot see the target.

(c) TIME ON TARGET. The observer tells the FDC when the rounds should impact.

(d) CONTINUOUS ILLUMINATION. When this is not already calculated by the FDC, the observer indicates the interval between rounds in seconds.

(e) COORDINATED ILLUMINATION. The observer tells the FDC to set the interval between illumination and HE shells.

(f) CEASE LOADING.

(g) CHECK FIRING. Halt immediately.

(h) CONTINUOUS FIRE. Load and fire as quickly as possible.

(i) REPEAT. Fire another round or more rounds, with or without adjustments.

(j) FOLLOWED BY. This is part of a phrase that indicates a change in the rate of fire, a change in the type of ammunition, or another order for fire for effect (for example [with WP being white phosphorus], WP FOLLOWED BY HE).

h. Correction of errors. When the FDC has made an error while reading backfire support data, the observer announces CORRECTION and transmits the correct data in their entirety.

i. Message to observer.

(1) Battery(ies) to fire for effect.

(2) Adjustment of battery.

(3) Changes to the initial call for fire.

(4) Number of rounds (per tube) to fire for effect.

(5) Target numbers.

(6) Additional information.

(a) Time of flight (moving target mission).

(b) Probable error in range (38 meters or greater [normal mission]).

(c) Angle-T (500 milliradians or greater).

j. Authentication. Challenge and reply.

3-11. Table 3-6 provides one sample call for fire transmission.

Table 3-6. Example of a call for fire transmission

GRID	
Observer	**Firing Unit**
F24, THIS IS J42, ADJUST FIRE, OVER.	J42, THIS IS F24, ADJUST FIRE, OUT.
GRID WM180513, ALTITUDE 300, OVER.	GRID WM180513, ALTITUDE 300, OUT.
INFANTRY PLATOON DUG IN, OVER.	INFANTRY PLATOON DUG IN, OUT.
	SHOT, OVER.
SHOT, OUT.	
	SPLASH, OVER.

Table 3-6. Example of a call for fire transmission (continued)

Observer (continued)	Firing Unit (continued)
SPLASH, OUT.	
END OF MISSION, 15 CASUALTIES, PLATOON DISPERSED, OVER.	END OF MISSION, 15 CASUALTIES, PLATOON DISPERSED, OUT.
SHIFT FROM KNOWN POINT	
Observer	**Firing Unit**
J42, THIS IS F24, ADJUST FIRE, SHIFT AB1001, OVER.	F24, THIS IS J42, ADJUST FIRE, SHIFT AB1001, OUT.
DIRECTION 2,420, RIGHT 400, ADD 400, OVER.	DIRECTION 2,420, RIGHT 400, ADD 400, OUT.
FIVE T-72 TANKS AT POL SITE, OVER.	FIVE T-72 TANKS AT POL SITE AUTHENTICATE JULIET NOVEMBER, OVER.
I AUTHENTICATE TANGO, OVER.	
	SHOT, OVER.
SHOT, OUT.	
END OF MISSION, TWO TANKS DESTROYED, THREE IN WOODLINE, OVER.	END OF MISSION, TWO TANKS DESTROYED, THREE IN WOODLINE, OUT.
POLAR	
Observer	**Firing Unit**
J42, THIS IS F24, ADJUST FIRE, POLAR, OVER.	F24, THIS IS J42, ADJUST FIRE, POLAR, OUT.
DIRECTION 2,300, DISTANCE 4,000, OVER.	DIRECTION 2,300, DISTANCE 4,000, OUT.
INFANTRY PLATOON DUG IN, OVER.	INFANTRY PLATOON DUG IN, OUT.
	SHOT, OVER.
SHOT, OUT.	
	SPLASH, OVER.
SPLASH, OUT.	
END OF MISSION, 15 CASUALTIES, PLATOON DISPERSED, OVER.	END OF MISSION, 15 CASUALTIES, PLATOON DISPERSED, OUT.
Legend: POL–petroleum, oils, and lubricants	

CLOSE AIR SUPPORT

3-12. There are two types of CAS requests: planned and immediate. The Army chain to corps processes planned requests for approval. The initiation of immediate requests occurs at any level, and the battalion operations officer (S-3), fire support officer, and air liaison officer process them.

3-13. Figure 3-3 on page 3-14 provides one sample call for CAS. Tables 3-7 and 3-8 on pages 3-15 through 3-17 detail CAS capabilities. Table 3-7 details fixed-wing aircraft CAS capabilities, and table 3-8 details unmanned aircraft system CAS capabilities. The format for requesting immediate CAS follows.

    a. Observer identification.

    b. WARNORD (request close air).

    c. Target description. At a minimum, include the type and number of targets, activity or movement, and point or area targets.

    d. Target location (grid) to include elevation.

    e. Desired time on target.

    f. Desired effects on target.

    g. Final control.

    h. Remarks. Report the following information.

        (1) Friendly locations.

        (2) Wind direction and hazards.

        (3) Threats such as air defense artillery or small arms.

19 September 2025   TC 3-21.76   3-13

```
JTAC: "       (aircraft call sign)       , this is    (JTAC/observer call sign)    "
"Type  (1,2,3)  control."

1. INITIAL POINT (IP): _____ NP459854 (or) X-ray _____

2. HEADING (IP TO TARGET): _____ 069 _____ MAGNETIC
                                 (OFFSET: LEFT/RIGHT)

3. DISTANCE (IP TO TARGET: _____ 9-8 _____ (NAUTICAL MILES)

4. TARGET ELEVATION: _____ 1140 _____ (FEET ABOVE MEAN SEA LEVEL)

5. TARGET DESCRIPTION: _____ 5 tanks attacking west _____

6. TARGET LOCATION: _____ NP675920 _____
              (UTM, LAT/LONG, VISUAL REFERENCES, AND OTHER DATE)

7. TYPE OF MARK: LASER CODE: 372

8. LOCATION OF FRIENDLIES: _____ 1000m SW of target _____

9. EGRESS: _____ NW to avoid artillery suppression _____

   (REMARKS) _____ --------- _____


   TIME ON TARGET (TOT): _____ --------- _____

   TIME TO TARGET (TTT) STANDBY: _____ --------- _____ PLUS: _____ --------- _____
                                      (MINUTES)           (SECONDS)
OMIT DATA NOT REQUIRED.
LINE NUMBERS ARE NOT TRANSMITTED.
ALL UNITS OF MEASURE ARE STANDARD.
SPECIFY IF OTHER UNITS OF MEASURE ARE BEING USED.
```

LEGEND

JTAC	JOINT TERMINAL ATTACK CONTROLLER	NW	NORTHWEST
LAT	LATITUDE	SW	SOUTHWEST
LONG	LONGITUDE	UTM	UNIVERSAL TRANSVERSE
m	METER		MERCATOR

Figure 3-3. Example of a close air support request

Table 3-7. Fixed-wing aircraft close air support capabilities

FIXED-WING AIRCRAFT CAPABILITIES AND COMMUNICATIONS EQUIPMENT						
Aircraft	Ordnance	Sensors and Markings	Data Link	Frequency Band	Frequency Hopping	COMSEC
F-15E Strike Eagle	JDAM, LJDAM, LGB, CBU/WCMD, AGM-130/158, JSOW, 20-mm cannon	Link 16	FLIR, SAR, Sniper, LITENING, Terrain-following radar	VHF-AM/FM, UHF, SATCOM	HQ II, SINCGARS	VINSON
AV-8B Harrier II	LGB, AGM-65-E, GP bombs, CBU, JDAM, LJDAM, 2.75-inch rockets, 5-inch Zuni	LITENING, CCD TV, FLIR, SAR, LUU-2/19	None	UHF, VHF-AM/FM	HQ II, SINCGARS	VINSON
A-10C Thunderbolt	LGB, AGM-65, GP bombs, CBU/WCMD, JDAM, 2.75-inch rockets, 30-mm cannon	LITENING, Sniper Pave Penny, Quickdraw, IZLID, LUU-2/19	SADL, VMF	UHFx2, VHF-AM/FM, SATCOM	HQ II, SINCGARS	VINSON
EC-130 Compass Call	Electronic attack	Link 16	SIGINT	3xUHF, 1xUHF, 2xHF, 2xSATCOM, IRC, NCCT chat, STE (AITG terminal also configurable to additional UHF, VHF, FM, SINCGARS, HQ, DAMA)	HQ II	

Table 3-7. Fixed-wing aircraft close air support capabilities (continued)

FIXED-WING AIRCRAFT CAPABILITIES AND COMMUNICATIONS EQUIPMENT						
Aircraft	Ordnance	Sensors and Markings	Data Link	Frequency Band	Frequency Hopping	COMSEC
AC-130J Ghostrider	105-mm howitzer, 40-mm cannon	IDS, PLS, LLLTV, beacon-tracking radar, IZLID, ATI, HIBEAM, LTD	Link 16	UHFx2, SATCOM, HFx2, VHF-AM/FMx3	HQ II, SINCGARS	VINSON, ANDVT
B-1B Lancer	JDAM (with pattern capability), GP bombs, CBU/WCMD, JASSM	DCI/JRE	SAR, sniper, GMTI/T, PSS-SOF	UHF, VHF, SATCOM voice, HF	HQ II, SINCGARS	VINSON

Legend: AITG–airborne integrated terminal group; AM–amplitude modulation; ANDVT–advanced narrowband digital voice terminal; ATI–ambient temperature illuminator; CBU–cluster bomb unit; CCD–charged-coupled device; COMSEC–communications security; DAMA–demand assigned multiple access; DCI–digital communications improvement; FLIR–forward-looking infrared; FM–frequency modulation; GMTI/T–ground moving target indicator and tracking; GP–general purpose; HF–high frequency; HIBEAM–high beam; HQ–HAVE QUICK [a frequency hopping system that protects military UHF radio traffic]; IDS–infrared detection set; IRC – internet relay chat; IZLID–infrared zoom laser illuminator designator; JASSM–joint air-to-surface standoff missile; JDAM–Joint Direct Attack Munition; JRE–Joint Range Extension; JSOW–joint standoff weapon; LGB–laser-guided bomb; LITENING–[a multi-sensor targeting and surveillance system]; LJDAM–laser JDAM; LLLTV–low light level television; LTD–laser target designator; mm–millimeter; NCCT–network centric collaborative targeting; PLS–precision locator system; PSS-SOF–Precision Strike Suite-Special Operations Forces; SADL–situation awareness data link; SAR–synthetic aperture radar; SATCOM–satellite communications; SIGINT–signals intelligence; SINCGARS–single-channel ground and airborne radio system; STE–secure telephone equipment; TV–television; UHF–ultrahigh frequency; VHF–very high frequency; VINSON–[an encrypted UHF/VHF communications system]; VMF–variable message format; WCMD–wind corrected munitions dispenser

Table 3-8. Unmanned aircraft close air support capabilities

UAS CAPABILITIES AND COMMUNICATIONS EQUIPMENT						
Aircraft	Ordnance	Sensors and Markings	Data Link	Frequency Band	Frequency Hopping	COMSEC
Hunter	None	MOSP (EO, IR, ELRF, LD), CRP	C-band (LOS)	UHF/VHF		
MQ-1B Predator	Hellfire	MTS, LTM, LTD	Link 16, PACWIND	UHF, VHF–AM/FM, SATCOM, VDL	HQ II, SINCGARS	
MQ-9 Reaper	Hellfire, GBU-12, GBU-38	MTS, SAR, GMTI, LTM, LTD	Link 16	UHF, VHF–AM/FM, SATCOM, VDL	HQ II, SINCGARS	VINSON
MQ-1C Gray Eagle	Hellfire	CSP (EO, IR, ELRF/LD), CRP, SAR/GMTI	TCDL (Ku band), SATCOM	VHF/UHF, SATCOM	HQ II	
RQ-7B Shadow	None	IR, EO, IRLP, LRF/D, LTD		VHF-FM, UHF2, UHF2-SATCOM2 (CRP is VHF-FM only)	HQ II, SINCGARS	VINSON
RQ-11B Raven	None	IR, EO, IR Pointer	None	None	None	None
Note: MQ and RQ are nomenclature designations for UASs.						
Legend: AM–amplitude modulation; C-band–[communications data link]; COMSEC–communications security; CRP–communications relay payload; CSP–common sensor payload; ELRF–eyesafe laser range finder; EO–electro-optical; FM–frequency modulation; GBU–guided bomb unit; GMTI–ground moving target indicator; HQ–HAVE QUICK [a frequency hopping system that protects military UHF radio traffic]; IR–infrared; IRLP–infrared laser pointer; LD–laser designator; LOS–line of sight; LRF/D–laser range finder/detector; LTD–laser target designator; LTM–laser target marker; MOSP–multi-mission optronic stabilized payload; MTS–multi-spectral targeting system; PACWIND–[an LOS full-motion video signal]; SAR–synthetic aperture radar; SATCOM–satellite communications; SINCGARS–single-channel ground and airborne radio system; TCDL–tactical common data link; UAS–unmanned aircraft system; UHF–ultrahigh frequency; VDL–video downlink; VHF–very high frequency; VINSON–[an encrypted UHF/VHF communications system]						

ARMY ATTACK AVIATION

3-14. Army attack aviation is defined as a hasty or deliberate attack in support of units engaged in close combat. During an attack, armed helicopters engage enemy units with direct fire that impacts nearby friendly forces. Targets range from a few hundred meters to a few thousand meters. A team, a platoon, or company-level ground unit Soldiers coordinate and direct Army attack aviation by using standardized Army attack aviation procedures in their units' SOPs. Table 3-9 provides one sample army attack aviation call for fire in its standard format with added notations for explanation.

Table 3-9. Army attack aviation call for fire format

ARMY ATTACK AVIATION CALL FOR FIRE FORMAT
1. Observer and warning order: **_LONGBOW 6_**, THIS IS **_OBSERVER 2_**, FIRE MISSION, OVER. (aircraft call sign)   (observer call sign)
2. Friendly location and mark: MY POSITION **_NP359654_**, MARKED BY **_STROBE_**. (TRP, grid, other)   (strobe, beacon, IR strobe, other)
3. Target location: TARGET LOCATION **_NP459854_**. (bearing [magnetic] and range [meters], TRP, grid, other)
4. Target description and mark: **_3 TANKS_**, MARKED BY **_INFRARED POINTER_**. (target description)   (infrared pointer, tracer, other)
5. Remarks: _____**_NONE_**_____ . (threats, danger close clearance, restriction, at my command, other)
NOTES: 1. Clearance. When the airspace is clear between the employing aircraft and the target, transmission of this brief <u>is</u> clearance to fire unless DANGER CLOSE or AT MY COMMAND is stated. 2. Danger close. For danger close fire, the observer or commander accepts responsibility for increased risk. State CLEARED DANGER CLOSE on line 5 and pass the initials of the on-scene ground commander. This clearance is sometimes preplanned. 3. At my command. For positive control of the aircraft, state AT MY COMMAND on line 5. The aircraft calls READY TO FIRE when ready.
Legend: IR–infrared; TRP–target reference point

3-15. During the planning process, the team-, platoon-, or company-level leaders are responsible for allowing enough time to conduct rehearsals between the ground unit and the aviation unit. Rehearsals ensure all participants in the mission have situational awareness of the plan, routes, capabilities, and limitations of each unit. They walk away understanding what they are to provide and how they are to support. Table 3-10 details rotary-wing aircraft CAS capabilities.

Table 3-10. Rotary-wing aircraft close air support capabilities

ROTARY-WING AIRCRAFT CAPABILITIES AND COMMUNICATIONS EQUIPMENT						
Aircraft	Ordnance	Sensors and Markings	Data Link	Frequency Band	Frequency Hopping	COMSEC
AH-6	7.62-mm MG, Caliber .50 MG, 2.75-inch rockets	FLIR, IR Pointer		VHF-AM/FMx2, UHF-AM/FM, SATCOM	HQ II, SINCGARS	VINSON, ANDVT
AH-64D/E	Hellfire (laser or RF), 2.75-inch rockets, 30-mm cannon	FLIR (LTD3), MMW, radar, DTV, IZLID		VHF-AM, VHF-FMx2	HQ II, SINCGARS	VINSON
Legend: AM–amplitude modulation; ANDVT–advanced narrowband digital voice terminal; COMSEC–communications security; DTV–day television; FLIR–forward-looking infrared; FM–frequency modulation; HQ–HAVE QUICK [a frequency hopping system that protects military UHF radio traffic]; IR–infrared; IZLID–infrared zoom laser illuminator designator; LTD–laser target designator; MG–machine gun; mm–millimeter; MMW–millimeter wave; RF–radio frequency; SATCOM–satellite communications; SINCGARS–single-channel ground and airborne radio system; UHF–ultrahigh frequency; VHF–very high frequency, VINSON–[an encrypted UHF/VHF communications system]						

This page intentionally left blank.

Chapter 4

Communications

The basic requirement of combat communications is to provide a rapid, reliable, and secure interchange of information. Communications are vital to mission success. This chapter helps the Ranger squad and platoon maintain effective communications and correct any radio antenna problems. It also discusses military radio communications equipment and automated net control devices.

EQUIPMENT

4-1. Paragraphs 4-1 through 4-4 discuss military radio communications equipment and automated net control devices. Each military radio has a receiver and transmitter. Rangers use several different types of radios. (See table 4-1.) Radios vary from high frequency, very high frequency, ultrahigh frequency, to tactical satellite.

Table 4-1. Characteristics of military radios

CHARACTERISTICS	AN/PRC-148	ASIP
Description	Multiband, interteam or intrateam radio	Multiband, multi-mission manpack
Frequency Range		
UHF	Yes	No
Power Output	Up to 5 watts	Up to 10 watts
Battery Requirements	Rechargeable lithium ion (included with the radio)	Any of BB-390, BB-2590, BB-590, BB-5590
Scanning	10 user-programmed nets (TACSAT or LOS frequencies)	Four channels in FM mode
Data Transmission		
LOS AM/FM	Yes	Yes
Commercial	Yes	No
Optional Internal	No	Yes
TACSAT	No	No
DAGR	No	No

Table 4-1. Characteristics of military radios (continued)

CHARACTERISTICS	AN/PRC-148	ASIP
	Optional Internal	
Dimensions and Weight	2.7 x 7.8 x 1.5 inches (6.9 x 19.8 x 3.8 centimeters)	3.4 x 5.3 x 10.2 inches (8.6 x 13.5 x 25.9 centimeters)
	2.2 pounds (1 kilogram) with battery	7.7 pounds (3.5 kilograms) without battery
Disadvantages	Lower power output than AN/PRC-117F(c) and AN/PRC-152	Lower power output than AN/PRC-117F(c), Limited frequency range, Inability to communicate with USAF aircraft
Immersion Depth	2 to 20 meters (7 to 66 feet)	1 meter (3 feet)
SA Reporting Capability	With remote control unit	With optional internal GPS
Special Features	Ship-to-shore, ground-to-ground, air-to-ground	Does not apply
Terrain Restrictions	LOS – open to slightly rolling terrain, TACSAT – any terrain	LOS – open to slightly rolling terrain
Legend: AM–amplitude modulation; ASIP–advanced system improvement program; DAGR–defense advanced GPS receiver; FM–frequency modulation; GPS–Global Positioning System; LOS–line of sight; SA–situational awareness; TACSAT–tactical satellite; UHF–ultrahigh frequency; USAF–United States Air Force; x–by		

4-2. Knowing each radio's capabilities is crucial in planning and requesting the most reliable and effective communications equipment for a particular mission. Military operations use four primary frequency ranges.

MANPACK RADIO ASSEMBLY (AN/PRC-119F)

4-3. To assemble a manpack radio, first check and install a battery. Next, follow these steps to position the antenna, set up the handset, set presets and frequencies, and scan.
   a. Inspect the battery box for dirt or damage.
   b. Stand the radio on its side with the battery cover facing up.
   c. Check the battery life condition (rechargeable BB-390 batteries).
   d. Place the battery in the box.
   e. Close and latch the battery cover.
   f. Return the radio to its upright position.
   g. If the battery is in used condition, enter the battery life condition into the radio.
      (1) Set FCTN (function) to LD (load).
      (2) Press BAT (battery) and then CLR (clear).
      (3) Enter the number recorded on the side of the battery.
      (4) Press STO (store).
      (5) Set FCTN to SQ ON (squelch on).

h. Inspect and position the antenna.
    (1) Inspect the whip antenna connector on the antenna and on the radio for damage.
    (2) Screw the whip antenna into the base.
    (3) Hand-tighten.
    (4) Carefully mate the antenna base with the RT ANT (radio transmitter antenna) connector.
    (5) Hand-tighten.
    (6) Position the antenna as necessary by bending the gooseneck.

Note. Keep the antenna as straight as possible. When the antenna is bent to a horizontal position, the radio likely requires turning before it can receive or transmit messages.

---

i. Set up the handset.
    (1) Inspect the handset for damage.
    (2) Push the handset connector onto the AUD/DATA (audio/data) connector and twist clockwise to lock in place.
j. Pack.
    (1) Place the RT (radio transmitter) in the field pack with the antenna on the left shoulder.
    (2) Fold the top flap of the field pack over the RT and secure the flap to the field pack using straps and buckles.
k. Set presets.
    (1) CHAN (channel): 1.
    (2) MODE: SC (single channel).
    (3) RF PWR (radio frequency power): HI (high).
    (4) VOL (volume): mid-range.
    (5) DIM (dimensions): full clockwise.
    (6) FCTN: LD.
    (7) DATA RATE: OFF.
l. Set single channel loading frequencies.
    (1) Obtain Ranger signal operating instructions.
    (2) Set FCTN: LD.
    (3) Set MODE: SC.
    (4) Set CHAN: MAN, CUE (manual cue). Or, set the channel (1 through 6) to that at which the frequency is stored.
    (5) Press FREQ (frequency). The display shows 000001, or the frequency RT is currently turned on.
    (6) Press CLR. The display shows five lines.
    (7) Enter the number of the new frequency. To correct a mistake, press CLR.
    (8) Press STO. The display blinks.
    (9) Set FCTN: SQ ON.
m. Clear frequencies.
    (1) Set MODE: SC.
    (2) Set CHAN: MAN, CUE. Or, set the channel (1 through 6) to that at which the frequency is to be stored.

       (3) Press FREQ.

       (4) Press CLR.

       (5) Press LOAD, STO.

       (6) Set FCTN: SQ ON.

   n. Scan multiple frequencies:

       (1) Load all desired frequencies using step l (Set single channel loading frequencies).

       (2) Set CHAN: CUE.

       (3) Set SC: FH (frequency hop).

       (4) Set FCTN: SQ ON.

       (5) Press STO. SCAN displays.

       (6) Press the number 8 to scan more than one frequency.

BASIC TROUBLESHOOTING

4-4. Basic troubleshooting skills are necessary for correcting the simple communications problems that occur during a mission. The ability to troubleshoot quickly often makes the difference between mission success and failure.

   a. <u>Check radio settings</u>.

       (1) Radio frequency: Load the proper frequency.

       (2) Set power output: HIGH.

       (3) Time when using frequency hop: Reset time.

       (4) Cryptographic (crypto) fill when using cipher text: Reload cryptographic fill from the automated net control device.

       (5) Position control knob: ON.

   b. <u>Check radio assembly and battery</u>.

       (1) Check the antenna fitting and attach a long whip or field-expedient antenna.

       (2) Check the hand-mike fitting for clean contacts and proper securement of the fitting to the radio.

       (3) Check the battery and install a fresh one as appropriate.

   c. With line of sight radios, moving to higher ground is often necessary to make radio contact, especially in densely vegetated or uneven terrain.

ANTENNAS

4-5. Paragraphs 4-5 through 4-21 discuss repair techniques, construction and adjustment, field-expedient antennas, antenna length and orientation, and improving marginal communications. Antennas are sometimes broken or damaged, causing communications degradation or failure. When a spare antenna is available, replace the bad one.

4-6. When there is no spare, the squad or platoon sometimes constructs an emergency antenna. The following suggestions are ways to repair antennas and antenna supports and to construct and adjust emergency antennas.

DANGER

Contact with the radiating antenna of a medium-power or high-power radio transmitter can result in personnel death or serious injury. <u>Turn off</u> the radio transmitter while adjusting the antenna.

WHIP AND WIRE ANTENNAS

4-7. When a whip antenna breaks in two, connect the broken part to the part attached to the base by joining the sections. To restore the antenna to its original length, add a piece of wire of nearly the same length as the missing part of the whip. Lash the pole support securely to both sections of the antenna. Thoroughly clean the sections before connecting them to the pole support to ensure good contact. Whenever possible, solder the connections.

4-8. The emergency repair of a wire antenna sometimes involves the repair or replacement of the wire serving as the antenna or transmission line. Other times, it involves the repair or replacement of the antenna support assembly. When wires of an antenna are broken, the antenna is repairable by reconnecting the broken wires. To do so, lower the antenna to the ground, clean the ends of the wires, and twist the wires together. Whenever possible, solder the connection.

4-9. When the antenna receives damage beyond repair, construct a new one. The lengths of the substitute antenna wires must be the same lengths as the originals. Antenna supports may also require repair or replacement in this situation. Any substitute serves for the damaged support, provided it is insulated and strong enough.

4-10. When the radiating element is not properly insulated, the field antennas have the potential to short to the ground and no longer work. Many common items make good field-expedient insulators. The best insulators are plastic or glass. Plastic spoons, buttons, bottle necks, and plastic bags make good insulators. Although wood and rope are less effective insulators, they are better than nothing. The radiating element (the antenna wire) touches only this supporting (nonconductive) insulator and the antenna terminal. Otherwise, it remains physically separate from everything else.

CONSTRUCTION AND ADJUSTMENT

4-11. There are specific methods for constructing and adjusting antennas. The best wire for antennas is made of copper or aluminum. However, in an emergency, any available wire serves the purpose. The exact length of most antennas is critical. The emergency antenna must be the same length as the original antenna.

4-12. Antennas usually survive heavy windstorms with tree trunks or strong branches to support them. To keep the antenna tight and from breaking or stretching when the trees sway, attach a spring or old inner tube to one end of the antenna. Another technique is to pass a rope through a pulley or eyehook, attach the rope to the end of the antenna, and heavily weight the rope to keep the antenna tight. To ensure the rope or wire guidelines do not interfere with the operation of the antenna, cut the wire into several short lengths and connect the pieces with insulators.

4-13. An improvised antenna often changes the performance of a radio set. A distant station is useful when testing the antenna. When the signal from this station is strong, this indicates the antenna is operating satisfactorily. When the signal is weak, adjustments to the height and length of the antenna and the transmission line maximize their capacity to receive the strongest signal at a given setting on the volume control of the receiver. This is the best method of tuning an antenna when transmission is dangerous or forbidden.

4-14. Some radio sets allow the use of the transmitter to adjust the antenna by setting the controls of the transmitter to normal and then tuning the system through adjustments to the antenna height and length and the transmission line length. This process obtains the best transmission output.

EXPEDIENT 292-TYPE ANTENNA

4-15. Originally developed for use in the jungle, these antennas improve communications with proper use. Their weight and bulk render them impractical for most squad or platoon operations. Units sometimes carry just the masthead and antenna sections and mount them onto wood poles or trees. Alternatively, units construct an expedient version with any insulated wire and other available material. For example, almost any plastic, glass, or rubber items serve as insulators. When these are unavailable, dry wood works.

4-16. The following steps lay out the process of constructing an expedient antenna.
  a. At the radio set, remove about 1 inch (2.5 centimeters) of insulation from each end of the wire.
  b. Connect the ends to the positive side of the cobra head connector such that the connections are tight or secure.
  c   Set up the correct frequency, turn on the set, and proceed with communications.

Note. Use the below planning considerations to determine the length of the elements (one radiating wire and three ground plane wires) for the desired frequency. Cut these elements from claymore mine wire or similar wire. The heavier the gauge, the better, but insulated copper core wire works best.

---

  d. Cut spacing sticks to the same length as the ground plane wires.
  e. Place the sticks in a triangle and tie their ends together with wire, tape, or rope.
  f. Attach an insulator to each corner and one end of each ground plane wire to each insulator.
  g. Bring the loose ends of the ground plane wires together, attach them to an insulator, and tie securely.
  h. Strip about 3 inches (7.5 centimeters) of insulation from each wire and twist them together.
  i. Tie one end of the radiating element wire to the other side of insulator and the other end to another insulator.
  j. Strip about 3 inches (7.5 centimeters) of insulation from the radiating element.
  k. Cut enough wire to reach from the proposed location of the antenna to the radio set.

Note. Keep this line as short as possible because excess length reduces the efficiency of the system.

---

  l   Tie a knot at each end to identify it as the hot lead.

    m. Remove insulation from the hot wire and tie it to the radiating element wire at insulator.

    n. Remove insulation from the other wire and attach it to the bare ground plane element wires at insulator.

    o. Tape all connections and do not allow the radiating element wire to touch the ground plane wires.

    p. Attach a rope to the insulator on the free end of the radiating element and toss the rope over the branches of a tree.

    q. Pull the antenna as high as possible, keeping the lead-in routed down through the triangle.

    r. Secure the rope to hold the antenna in place.

ANTENNA LENGTH-PLANNING CONSIDERATIONS

4-17. The length of an antenna is considered in the construction of field expedients. At a minimum, a calculation of the appropriate length of a field-expedient antenna factors a quarter of the frequency wavelength. Another important factor in line of sight communications is the height of the antenna in relation to the receiving station. The higher the antenna, the greater the range the radio transmission has.

4-18. Terrain and curvature of the Earth affect line of sight communications by absorbing very high frequency and ultrahigh frequency communications into the Earth's surface. This is overcome by increasing antenna height, power output, and radio frequency. Since radio frequencies are predesignated and power output is limited to the capabilities of the radio set, manipulating the two variables antenna length and height tends to increase radio communication range. Using the following formulas, it is possible to plan for the use of field-expedient antennas, determine the best location to gain and maintain communication, and plan for communication windows as necessary.

4-19. The following equation calculates the physical length of an antenna in feet. It gives the antenna length in feet for a 1/4-wavelength of the frequency. To determine the antenna length in feet for a full wavelength antenna, multiply the antenna length by 4.

$X = 234/\text{frequency (freq)}$

(X = the length of the antenna in feet; freq = the radio frequency used)

   EXAMPLE:

      $234/38.950 = 6.01$ feet (1/4-wavelength antenna)

      6.01 feet $\times 2 = 12.02$ feet (1/2-wavelength antenna)

      6.01 feet $\times 4 = 24.04$ feet (full-wavelength antenna)

4-20. The curvature of the Earth allows a person standing 5 feet, 7 inches (170 centimeters) tall and looking across a flat surface to see objects approximately 4.7 kilometers (2.9 miles) in the distance. (See figure 4-1 on page 4-8.) Anything beyond this is below the horizon (dead space). To see beyond that distance, the person moves to a higher elevation.

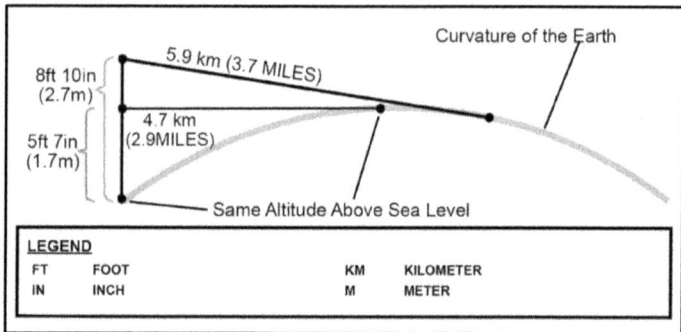

Figure 4-1. Curvature of the Earth

4-21. Line of sight communications are subject to this same principle. For this form of communication, when on low ground such as in a valley, draw, or depression, antenna height is taller, and when on high ground, antenna height is shorter. The following formula calculates the requisite antenna height for a given distance and computes an appropriate antenna height to compensate for the curvature of the Earth.

X = 234/frequency (freq)

(X = the length of the antenna in feet; freq = the radio frequency used)

Distance in kilometers (dkm) from receiving station = square root of (12.7 x Am) where Am is the antenna height in meters

EXAMPLE:

Known height: dkm = √(12.7 x Am)

dkm = √(12.7 x 1.7m)

or

Unknown height: Am = 0.07874 x (dkm)²

Am = 0.07874 x (4.7km)²

Chapter 5

Demolitions

This chapter introduces Rangers to the characteristics of explosives, initiation systems, modernized demolition initiator (MDI) components, detonation systems, and safety considerations. (See TM 3-34.82 for information on demolition.)

EXPLOSIVES

5-1. Low explosives have a detonating velocity of up to 1,300 feet per second, which produces a pushing or shoving effect. HEs have a detonating velocity of 3,280 to 27,888 feet per second, which produces a shattering effect. Table 5-1 provides some information on explosives of common use to Rangers for demolition.

Table 5-1. Characteristics of U.S. demolition explosives

NAME	APPLICATIONS	DETONATION VELOCITY		RE FACTOR	FUME TOXICITY	WATER RESISTANCE
		M/S	FT/S			
PETN	Det cord blasting caps demolition charges	8,300	27,200	1.66	Slight	Excellent
RDX	Blasting caps composition explosive	8,350	27,400	1.60	Moderate	Excellent
Composition C-4 (M112)	Cutting and breaching charges	8,040	26,400	1.34	Slight	Excellent
Det cord	Priming demolition charge	6,100 to 7,300	20,000 to 24,000	1.66	Slight	Excellent
Note: TNT = 1.00 relative effectiveness						
Legend: det cord–detonating cord; FT/S–feet per second; M/S–meters per seconds; PETN–pentaerythritol tetranitrate; RDX–Royal Demolition Explosive (hexahydrotrinitrotriazine); RE–relative effectiveness; TNT–trinitrotoluene						

INITIATING (PRIMING) SYSTEMS

5-2. The best way to prime demolition systems is with MDIs. These are blasting caps attached to various lengths of time fuze or shock tube. (See figure 5-1 on page 5-3 for a technique for cutting shock tube to length.) Using MDIs with a fuze igniter and detonating cord creates many firing systems. In the absence of MDIs, field-expedient methods are useful.

5-3. A shock tube is a thin, plastic tube of extruded polymer with a layer of special explosive material on the interior surface. Explosive material propagates a detonation wave that moves along the shock tube to a factory-crimped and -sealed blasting cap. Detonation is normally contained within the plastic tubing. However, shock tubes can burn skin when hand-held. Shock tubes have a few notable advantages:

- They are extremely reliable.
- They offer instant electric initiation, and they also prevent radio transmitters, static electricity, and the like from accidentally causing an initiation.
- They are extendable by using leftover sections from previous operations.

5-4. Five types of MDI blasting caps are available to replace the M6 electric and M7 nonelectric blasting cap. Three are high strength, and two are low strength. High strength blasting caps prime all standard military explosives including detonating cord and initiate the shock tube for other MDI blasting caps. Details regarding the five types follow.

 a. M11 (high strength).
  (1) Factory-crimped to 30 feet (about 9 meters) of shock tube.
  (2) An attached, movable J-hook for quick and easy attachment to a detonating cord.
  (3) A red flag attached 1 meter (3 feet) from the blasting cap and a yellow flag attached 2 meters (6.5 feet) from the blasting cap.
 b. M14 (high strength).
  (1) Factory-crimped to 7.5 feet (about 2.25 meters) of time fuze.
  (2) Initiation possible by a fuze igniter or match.
  (3) About a 5-minute burn time for the total length.
  (4) Yellow band indicators of calibrated 1-minute time intervals.
 c. M15 (high strength).
  (1) Two blasting caps factory-crimped to 70 feet (21.3 meters) of shock tube.
  (2) Factory-crimped to 7.5 feet (about 2.25 meters) of time fuze.
  (3) Delay elements that allow for staged detonations.
  (4) Low strength blasting caps useful as relay devices to transmit a shock tube detonation impulse from an initiator to a high strength blasting cap.
 d. M12 (low strength): factory-crimped to 500 feet (about 152.5 meters) of shock tube on a cardboard spool.
 e. M13 (low strength): factory-crimped to 1,000 feet (304.8 meters) of shock tube.

Note. High altitudes and colder temperatures increase burn time.

CUT SHOCK TUBE WITH A SHARP KNIFE OR
SCISSORS. ENSURE YOU HAVE A CLEAN CUT.

Figure 5-1. Technique for cutting shock tube

DETONATION (FIRING) SYSTEMS

5-5. The two types of firing systems are MDI alone or MDI plus a detonating cord. An MDI-alone firing system uses MDI components to construct the initiation set, transmission, and branch lines, and MDI blasting caps prime the explosive charges. Figure 5-2 on page 5-4 depicts some common firing system components. The following steps lay out the process of constructing the charge.

   a. Emplace and secure an explosive charge such as C4, trinitrotoluene, or a cratering charge on the target.
   b. Place a sandbag or other easily identifiable marker over the M11, M14, or M15 blasting cap.
   c. Connect to an M12 or M13 transmission line as desired.
   d. Connect a blasting cap with shock tube to an M14 cap with time fuze. (See figure 5-3 on page 5-5 for a technique for attaching shock tube, in this case to M81s.) Cut the time blasting fuze to the desired delay time.
   e. Prime the explosive charge by inserting the blasting cap into the charge.
   f. Visually inspect the firing system for possible misfire indicators such as cracks, bulges, or corrosion.
   g. Return to the firing point and secure a fuze igniter to the cut end of the time fuze. Figure 5-4 on page 5-6 depicts a dual initiating system, and figure 5-5 on page 5-7 depicts a British J initiating system.
   h. Remove the safety cotter pin from the igniter's body.
   i. Actuate the charge by grasping the igniter body with one hand while sharply pulling the pull ring.

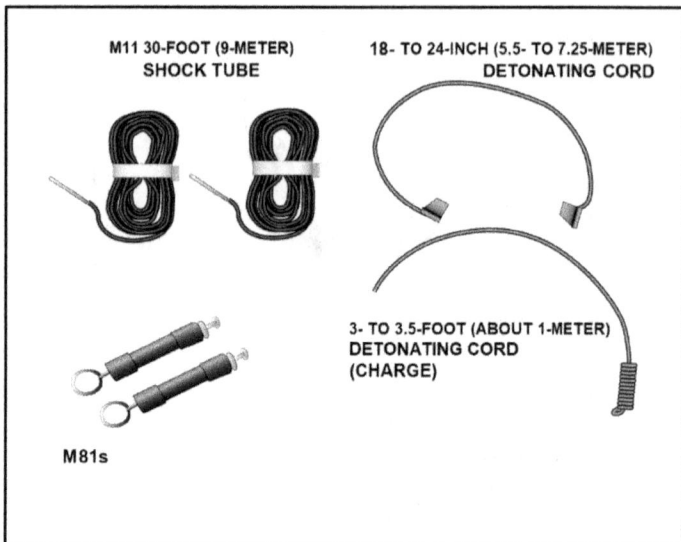

Figure 5-2. Firing system components

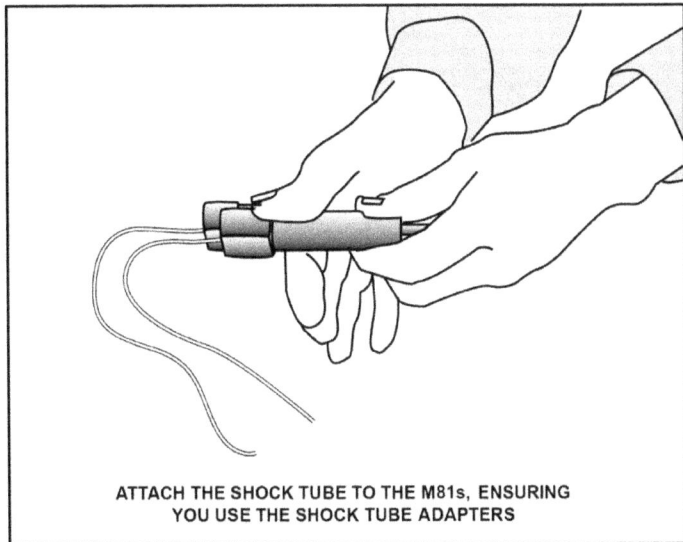

ATTACH THE SHOCK TUBE TO THE M81s, ENSURING YOU USE THE SHOCK TUBE ADAPTERS

Figure 5-3. Technique for attaching shock tube to M81s

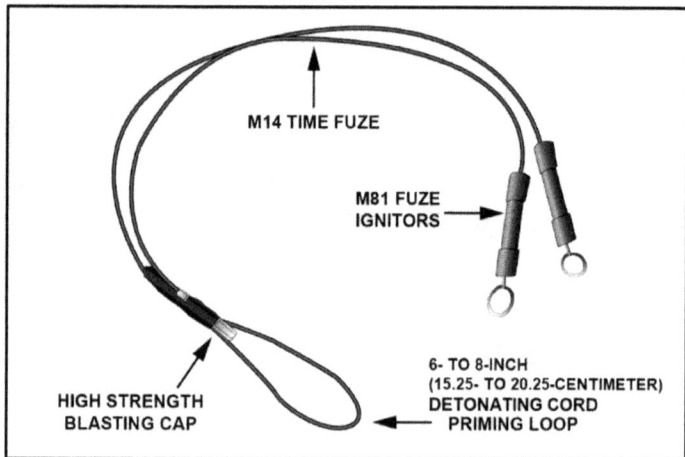

Figure 5-4. Dual initiating system

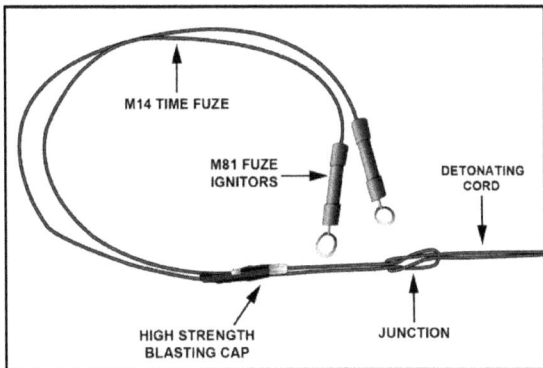

Figure 5-5. British J initiating system

CHARGING

5-6. The demolitions effects simulator is the only charge constructed in Ranger School. Figure 5-6 on page 5-8 depicts a basic constructed demolitions effects simulator charge system, and figure 5-7 on page 5-8 depicts a demolitions effects simulator charge with a dual initiating system. Figures 5-8 and 5-9, on pages 5-9 and 5-10 respectively, provide examples of material usage in the construction. Demolitions effects simulator charges share some common components:

- Two MN08/M81 fuze ignitors.
- Two MN06/M14 time fuzes.
- 3 feet of M456 detonating cord type-1.
- Cardboard cutout.

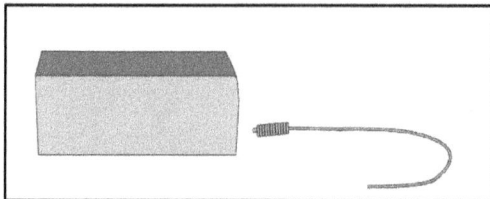

Figure 5-6. Demolitions effects simulator charge – constructed system

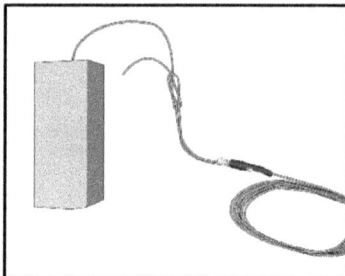

Figure 5-7. Dual initiated demolitions effects simulator charge

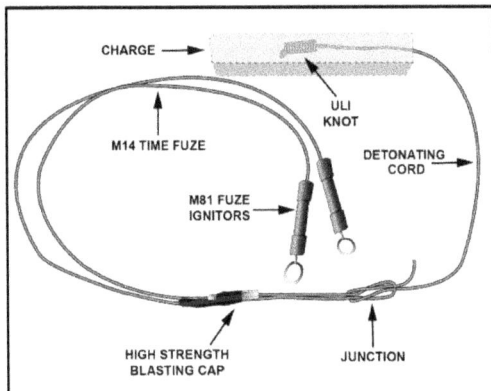

Figure 5-8. Demolitions effects simulator charge with uli knot and 12 to 18 inches of detonating cord

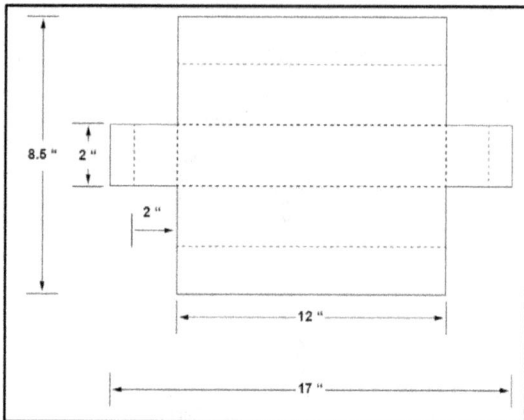

Figure 5-9. Demolitions effects simulator charge cardboard cutout dimensions

SAFETY

5-7. MDI is not recommended for belowground use except in quarry operations with water-gel or slurry explosives. Detonating cord is useful when it is necessary to bury primed charges. The M15 blasting cap is not for use with M1 dynamite. The M15 delay blasting cap is for use solely with water-gel or slurry explosives.

5-8. Do not handle misfires downrange until the requisite 30-minute waiting period elapses for both primary and secondary initiation systems and until the completion of all other safety precautions.

5-9. Never yank or pull hard on the shock tube. Doing so can actuate the blasting cap. Do not dispose of used shock tubes by burning due to the potential release of toxic fumes from burning plastic.

5-10. Always use protective equipment when handling demolitions. Minimum protection consists of leather gloves, ballistic eye protection, and a helmet.

5-11. All initiation systems and demolition charges have markings of the date of construction, minimum safe distance, and name of the person who constructed the charge.

5-12. When cutting detonating cord or time fuze, place a piece of 100 miles per hour (mph) tape over the cut end to prevent the explosive from falling out or moisture getting in, both of which can cause a misfire.

Chapter 6

Movement

So that Rangers can survive in the battlespace, their leaders enforce stealth, dispersion, and security in all tactical movements. The leader is skilled in all movement techniques. (See ATP 3-21.8 for information on movement techniques.)

MOVEMENT FORMATIONS

6-1. Movement formations include elements and Rangers arranged in relation to each other. Fire teams, squads, and platoons use several formations. Formations give the leader control based on an METT-TC (I) analysis. Leaders position themselves where they can best command and direct the formations, which figure 6-1 depicts. Typical formations are the line, vee, echelon, diamond, wedge, and file.

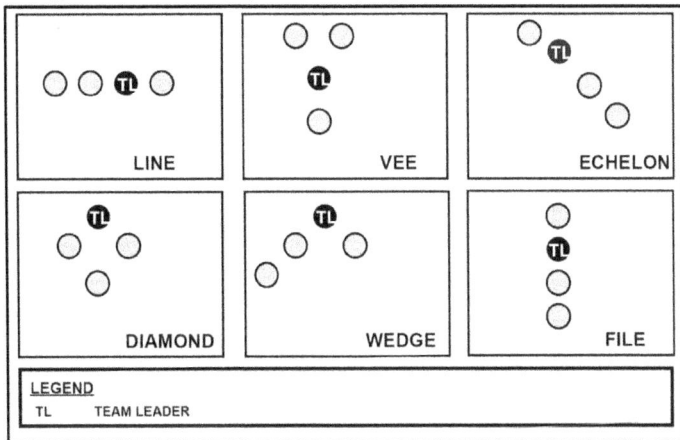

Figure 6-1. Formation types

6-2. Formations allow the fire TL to lead by example. (Follow me and do as I do!) The leader remains visible to all the Rangers in the team.

6-3. Squad formations reflect fire team formations. Squad formations are very similar, but with more Rangers. Squads operate in lines and files similarly to fire teams. When squads operate in wedges or echelon, the fire teams use those formations and simply arrange themselves in a column or with one team behind the other. Squads also use the vee, wherein one team forms the lines of the vee with the SL at the front (or the point of the vee) for mission command. Platoons use the same formations as squads. When the unit operates as a platoon, the PL carefully selects the location for the machine guns in the movement formation.

MOVEMENT TECHNIQUES

6-4. The selection of a movement technique is based on the likelihood of enemy contact and the relative need for speed. Specifically, the factors to consider include control, dispersion, speed, and security. Movement techniques are neither fixed nor are they formations. Instead, movement techniques are distinguished by a set of criteria such as the distances between individual Rangers and between teams or squads.

6-5. Movement techniques are not fixed formations. The distances between Soldiers, teams, and squads vary based on mission, enemy, terrain, visibility, and any other factor that affects control. There are three movement techniques: traveling, traveling overwatch, and bounding overwatch. (See table 6-1.) Individual movement techniques include high and low crawl and 3- to 5-second rushes from one covered position to another.

Table 6-1. Characteristics of movement techniques

MOVEMENT TECHNIQUES	WHEN NORMALLY USED	CHARACTERISTICS			
		Control	Dispersion	Speed	Security
Traveling	Contact not likely	More	Less	Fast	Less
Traveling Overwatch	Contact possible	Less	More	Slow	More
Bounding Overwatch	Contact expected	Most	Most	Slowest	Most

6-6. These movement techniques enable leaders to conduct actions on contact and tactical mission tasks and to make natural transitions to fire and movement. When analyzing the situation, leaders know some enemy positions. However, most of the time, enemy positions exist as templated positions only. Templated positions are the leader's best guess from analyzing the terrain and an existing knowledge of the enemy. Throughout the operation, leaders continuously try to confirm or deny both the known positions as well as the likely positions.

6-7. Movement techniques vary depending on the METT-TC (I) variables. What remains constant is the visibility of a fire TL to all the Rangers in the team. Likewise, the lead SL remains visible to the PL. Leaders control movement with hand and arm signals, and they use radios only when necessary. Leaders match the movement technique to the situation as follows.

    a. Traveling. Enemy contact is not likely, but speed is necessary. The target distance between Rangers is 10 meters (33 feet) and 20 meters (66 feet) between squads. Some characteristics are: (1) more control than traveling overwatch but less control than bounding overwatch; (2) minimum dispersion; (3) maximum speed; (4) minimum security.

    b. Traveling overwatch. Enemy contact is possible. This is the most often used movement technique. The target distance between Rangers is 20 meters (66 feet) and 50 meters (164 feet) between teams. Leaders keep in mind the following considerations for this technique.

(1) Only the lead squad uses traveling overwatch. However, in cases targeting greater dispersion, all squads may use it.

(2) In other formations, all squads use traveling overwatch unless the PL specifies not to do so. Traveling overwatch offers good control, dispersion, speed, and security forward.

(3) The lead squad is far enough ahead of the rest of the platoon to detect or engage any enemy before the enemy observes or fires on the main body. However, the lead squad stays between 50 and 100 meters (164 to 328 feet) in front of the platoon to enable the platoon's support with small arms fire. The distance depends on terrain, vegetation, and light and weather conditions.

c. Bounding overwatch. Enemy contact is likely or the unit is crossing a danger area. Both squads and platoons have bounding and overwatch elements. The bounding element moves while the other one occupies a position from which it overwatches by fire the bounding element's route. The bounding element remains within firing range of the overwatching element at all times. Leaders keep in mind the following considerations for this technique.

(1) Characteristics. Bounding overwatch offers maximum control, dispersion, and security with minimum speed.

(2) Types of bounds.

(a) Successive bounds. One element moves to a position, and then, the overwatching element moves to a position generally on the same line with the first element.

(b) Alternating bounds. One element moves into position, and then, the overwatching element moves to a position in front of the first element.

(3) Length. The length of a bound depends on the terrain, visibility, and control.

(4) Instructions. Before a bound, the leader gives the following instructions to subordinates.

(a) Direction of the enemy if known.

(b) Position of the overwatch elements.

(c) Next overwatch position.

(d) Route of the bounding element.

(e) What to do after the bounding element reaches the next position.

(f) How the elements receive follow-on orders.

(5) Squad bounding overwatch (see figure 6-2 on page 6-4). Rangers leave about 20 meters (66 feet) between themselves. The distance between teams and squads varies.

(6) Platoon bounding overwatch. When platoons use bounding overwatch (see figure 6-3 on page 6-5), one squad bounds, a second squad overwatches, and a third awaits orders. Rangers leave about 20 meters (66 feet) between them. The distance between teams and squads varies. FOs stay with the overwatching squad to call for fire. PLs normally stay with the overwatching squad, which uses machine guns and attached weapons to support the bounding squad. Another technique is to have one squad use bounding overwatch while the other two use traveling or traveling overwatch. The following considerations inform a leader's decision on where to move the bounding element.

(a) Enemy's likely action.

(b) Mission.

(c) Routes to the next overwatch position.

(d) Weapon ranges of the overwatching unit.

(e) Responsiveness of the rest of the unit.

(f) Fields of fire at the next overwatch position.

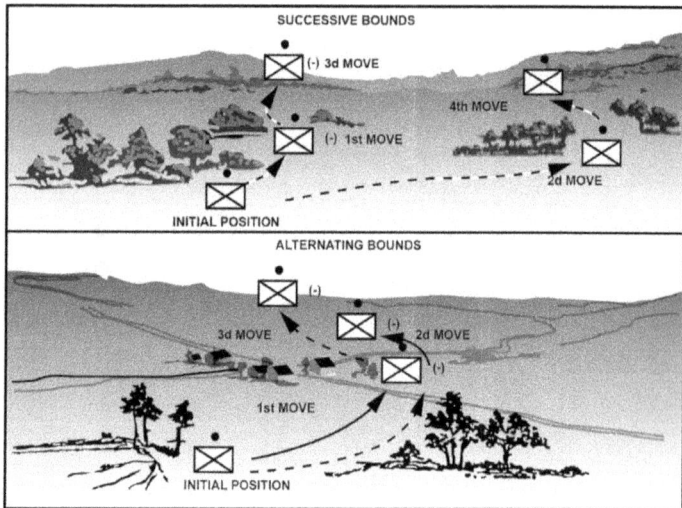

Figure 6-2. Example of squad bounding overwatch

TC 3-21.76

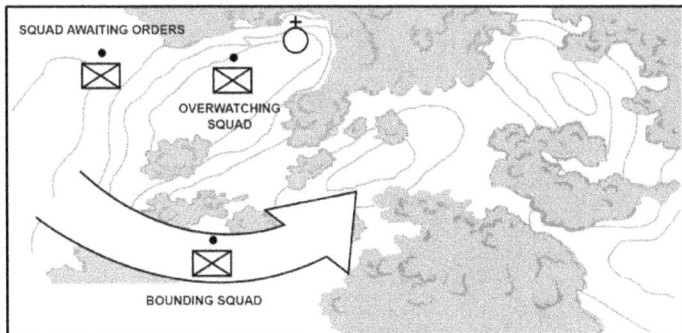

Figure 6-3. Example of platoon bounding overwatch

STANDARDS

6-8. A unit moves on a designated route or arrives at a specified location in accordance with the OPORD, maintaining accountability of all assigned and attached personnel. A unit uses the movement formation and technique ordered by the leader, which is based on the METT-TC (I) variables. Leaders remain oriented (within 200 meters [219 yards]) and follow a planned route unless METT-TC (I) variables dictate otherwise.

6-9. During movement, the unit maintains 360-degree security and remains 100-percent alert. During halts, the unit maintains 360-degree security and remains at least 75-percent alert. Whenever the unit makes contact with the enemy, it does so with the smallest element possible. The unit uses control measures during movement, such as headcounts, rally points, or phase lines.

FUNDAMENTALS

6-10. Mission accomplishment depends on successful land navigation. The patrol uses stealth and vigilance to avoid chance contact. Each patrol has designated primary and alternate compass and pace Soldiers. All leaders except fire TLs move inside their formations for best control over the platoon.

Note. The point person, whose sole responsibility is forward security for the element, never performs the compass or pace duties

---

6-11. Patrols use stealth and cover and concealment of the terrain to their maximum advantages. Whenever possible, the patrol moves during limited visibility to maximize the technological advantages of night-vision devices (NVDs) and hinder the enemy's ability to detect the patrol. They exploit the enemy's weaknesses and try to time movements to coincide with other operations that distract the enemy.

6-12. The patrol continues to use active and passive security measures. The leader assigns subunit responsibilities for security in danger areas, PBs, and the objective area. The leader plans fire support such as mortars, artillery, tactical air, attack helicopter, and naval gunfire.

6-13. The enemy threat and terrain determine which one of the three movement techniques is appropriate to use. Fire teams maintain visual contact, but enough distance remains between them that the entire patrol would not become engaged if it made contact. Fire teams spread their formations appropriately to gain better observation to the flanks. Though widely spaced, Rangers retain their relative positions in their wedge formations and follow their TLs. Only extreme situations call for the file formation. The lead squad secures the front and is responsible for navigation. For a long movement, the PL may rotate lead squad responsibilities. The fire team or squad in the rear is charged with rear security. Leaders vary movement techniques to meet the changing situation.

6-14. The patrol achieves 360-degree security, high and low. Within a fire team, squad, and so on, the leader assigns appropriate sectors of fire to subordinates. This ensures the battlespace is covered, including trees, multistoried structures, tunnels, sewers, and ditches.

TACTICAL MARCHES

6-15. Platoons conduct two types of marches with the company: foot marches and motor (road) marches. A foot march is successful when troops arrive at the destination at the prescribed time, physically able to execute their tactical mission.

6-16. To meet the standard, the unit crosses the start point and RP at the times specified in the order and follows the prescribed route, rate of march, and interval without deviation unless enemy or higher HQ actions require it. The fundamentals of tactical marches include effective control, detailed planning, and rehearsals. Leaders keep in mind the following considerations.
   a. METT-TC (I) variables.
      (1) Mission—task and purpose.
      (2) Enemy—intentions, capabilities, and COA(s).
      (3) Terrain and weather—road condition, trafficability, and visibility.
      (4) Troops and support—condition of Rangers and their loads and the number and types of weapons and radios.
      (5) Time—start time, release time, rate of march, and time available.
      (6) Civilians—movement through populated areas, refugees, and operations security.
      (7) Information—synergize all the variables to create actionable intelligence.
   b. Task organization.
      (1) Security—advance and trail teams.
      (2) Main body—the two remaining line squads and weapons squad.
      (3) HQ mission command.
      (4) Control measures.
   c. Control measures.

    (1) Start point and RP—given by higher HQ.
    (2) Checkpoints. At checkpoints, report to higher HQ to remain oriented.
    (3) Rally or rendezvous points—useful when elements become separated.
    (4) Locations of leaders—where they can best control their elements.
    (5) Communications plan—locations of radios, frequencies, call signs, and operation schedules.
    (6) Dispersion between Rangers.
        (a) 3 to 5 meters (10 to 16 feet) by day.
        (b) 1 to 3 meters (3 to 10 feet) by night.

  d. A march order is issued as an OPORD or FRAGORD or as an annex to either type of order. (An operational overlay or strip map is useful for this.) The march order includes the following information.
    (1) Formations and order of movement.
    (2) Route of march, assembly area, start point, RP, rally points, checkpoints, and break or halt points.
    (3) Start point time, RP time, and rate of march.
    (4) March interval for squads, teams, and individuals.
    (5) Actions on enemy contact—air and ground.
    (6) Actions at halts.
    (7) Detailed plan of fire support for the march.
    (8) Water supply plan.
    (9) Medical evacuation (MEDEVAC) plan.

6-17. Platoons comprise several Rangers fulfilling different positions and responsibilities. They include the PL, PSG, SLs, security squad, assault squads, medic, and individuals. The details of their duties before, during, and after the movement follow.

  a. PL.
    (1) Before: issues WARNORD, OPORD, or FRAGORD; inspects and supervises movement preparations.
    (2) During: ensures the unit makes its movement time, maintains the interval, and remains oriented; maintains security; checks the condition of the Rangers; spot-checks water discipline and field sanitation.
    (3) After: ensures the Rangers are prepared to accomplish their mission; supervises SLs; ensures the Rangers receive any necessary medical coverage.

  b. PSG.
    (1) Before: helps the PL; makes recommendations; enforces uniform and packing lists; obtains accountability of the Rangers prior to the start point time.
    (2) During: controls stragglers; assists the PL in maintaining the proper interval and security.
    (3) At halts: maintains accountability; enforces security and the welfare of the Soldiers; enforces field sanitation and litter discipline; performs preventive medicine; confirms the headcount prior to leaving the halt.
    (4) After: coordinates for water, rations, and medical supplies; recovers any casualties.

  c. SLs.
    (1) Before: provide detailed instruction to the TLs; inspect boots and socks for serviceability and proper fit; check for adjustment of equipment, full canteens, and equal distribution of loads.
    (2) During: control squad; maintain the proper interval between the Rangers and the equipment; enforce security; remain oriented.

(3) At halts: ensure the maintaining of security; provide the Rangers with water resupply as detailed; physically check the squad members; ensure the squad drinks water and changes socks as appropriate and rotates heavy equipment. (Units plan the latter in detail to avoid confusion before, during, and after halts.)

(4) After: occupy squad sector assembly area, conduct a foot inspection and report the condition of the Soldiers to the PL; prepare them to accomplish the mission.

d. Security squad.

(1) Serves as point element for platoon.

(2) Reconnoiters the route to the start point.

(3) Calls in checkpoints and provides frontal security and early warning.

(4) Maintains rate of march, moving 10 to 20 meters (33 to 66 feet) in front of the main body.

(5) The trail team provides rear security, moving 10 to 20 meters (33 to 66 feet) behind the main body.

e. Assault squads.

(1) Provide left, right, and rear security for the platoon.

(2) Prepare to provide additional combat power to the security squad in the scenario of their receiving contact and move 10 to 20 meters (33 to 66 feet) behind the security squad.

f. FO/radio-telephone operator (RTO).

(1) Maintains constant communication with higher HQ, moves with the PL, and sends all reports.

(2) Builds any necessary field-expedient antennas.

g. Medic.

(1) Assesses and treats march casualties.

(2) Advises the chain of command on the evacuation and transportation of casualties.

h. Individuals.

(1) Maintain the interval and follow the TL's example.

(2) Relay hand and arm signals.

(3) Remain alert during movement and at halts.

MOVEMENT DURING LIMITED VISIBILITY CONDITIONS

6-18. During hours of limited visibility, the platoon uses surveillance, target acquisition, and night observation devices to enhance effectiveness. Leaders retain their abilities to control, navigate, maintain security, and move during limited visibility.

6-19. When visibility is poor, methods that aid in control include moving leaders closer to the front, reducing the platoon speed, and using luminescent tape on equipment. Leaders also reduce the intervals between Soldiers and elements and execute headcounts.

6-20. While navigating during limited visibility, the unit uses the same techniques as in daylight, but leaders exercise more care to keep the patrol oriented. To maintain security, leaders enforce strict noise and light disciplines, and units use radio-listening silence and camouflage. Using the terrain to avoid detection by enemy surveillance or NVDs, Rangers make frequent halts to stop, look, listen, and smell. Whenever possible, they mask the sounds of movement. Rain, wind, and flowing water disguise movement sounds very efficiently.

6-21. Leaders plan detail actions to be taken at rally points. All elements maintain communications at all times. There are two techniques for actions at rally points:

- Minimum force — Patrol members assemble at the rally point, and the senior leader assumes command. Upon the assembly and organization of the minimum force (designated in the OPORD), the patrol continues the mission.
- Time available — The senior leader determines whether the patrol has enough time remaining to accomplish the mission.

6-22. During halts, the unit posts security and covers all approaches into the sector with key weapons. Leaders use two types of halts with Rangers:

- Short halt — This typically takes 1 to 2 minute(s). Rangers seek immediate cover and concealment and take a knee. Leaders assign sectors of fire.
- Long halt — This typically takes more than 2 minutes. Rangers assume the prone position behind cover and concealment. Leaders ensure Rangers have clear fields of fire and assign sectors of fire.

DANGER AREAS

6-23. A danger area is any place on a unit's route where the leader determines the potential exposure of the unit to enemy observation or fire. Some examples of danger areas are open areas, roads and trails, urban terrain, enemy positions, and natural and artificial obstacles. Units bypass danger areas whenever possible.

6-24. Standards, fundamentals, and techniques exist for crossing danger areas. Rangers take the following steps
  a. Standards.
    (1) Prevent the enemy from surprising the main body.
    (2) Move all personnel and equipment across the danger area.
    (3) Prevent decisive engagement by the enemy.
  b. Fundamentals.
    (1) Designate near side and far side rally points.
    (2) Secure near side, left and right flank, and rear security.
    (3) Reconnoiter and secure the far side.
    (4) Cross the danger area.
    (5) Plan for fires on all known danger areas.
  c. Techniques for crossing danger areas.
    (1) Linear danger area actions for a squad (see figure 6-4 on page 6-10).
      (a) The Alpha TL observes the linear danger area and sends the hand and arm signal to the SL, who determines to bound across.
      (b) The SL directs the Alpha TL to move the team across the linear danger area far enough to fit the remainder of the squad on the far side of the linear danger area. The Bravo team moves to the linear danger area to the right or left to provide an overwatch position prior to the Alpha team's crossing.
      (c) The SL receives the hand and arm signal that it is safe to move the rest of the squad across. (The Bravo team is still providing overwatch.)
      (d) The SL moves with the RTO and Bravo team across the linear danger area. (A team provides overwatch for squad missions.)
      (e) The A team assumes original azimuth at the SL's command or by hand and arm signal.

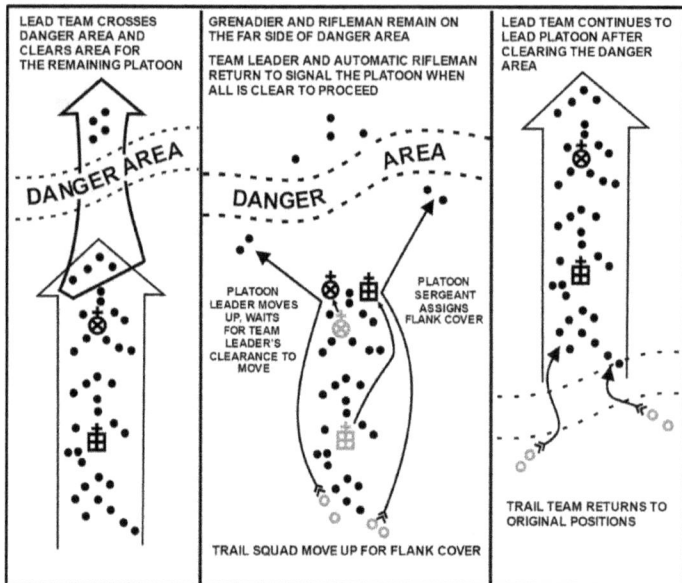

Figure 6-4. Examples of linear danger area actions

(2) <u>Linear danger area actions for a platoon</u>.
    (a) The lead squad halts the platoon and signals DANGER AREA.
    (b) The PL moves forward to the lead squad to confirm the danger area and then decides whether the current location is suitable for crossing.
    (c) The PL confirms the danger area or crossing site and establishes near side and far side rally points.
    (d) On the PL's signal, the trail squad moves forward to establish left and right near side security.

    (e) Upon near side security's establishment, the Alpha team of the lead squad, with the SL, moves across to confirm there is enough room to fit the rest of the platoon on the far side of the linear danger area.

    (f) After stopping to conduct look, listen, and smell observations, the SL signals ALL CLEAR to the PL. During daylight, a hand signal such as a thumbs up serves well as this signal. At night, a tool such as an infrared or a red lens serves as a better signal.

    (g) The PL then directs the Bravo team of the lead squad to bound across by team, link up with the Alpha team of the lead squad, and pick up a half step while the rest of the platoon crosses.

    (h) The PL then crosses with the RTO, FO, WSL, and one gun team.

    (i) Once across, the PL signals the second squad in movement to cross.

    (j) The PSG with the medic and one gun team crosses after the second squad is across, sterilizing the central crossing site.

    (k) The PSG signals the security squad to cross at their location.

    (l) The PSG calls the PL via frequency modulation (FM) radio to confirm all elements are across.

    (m) The PL directs the lead squad to pick up normal rate of movement.

Note. The PL plans fires on all known larger danger area crossing sites. Near side security in overwatch sterilizes signs of the patrol.

---

    (3) Danger area—small or open—actions (see figure 6-5 on page 6-12)

      (a) The lead squad halts the platoon and signals DANGER AREA.

      (b) The PL moves forward to the lead squad to confirm the danger area.

      (c) The PL confirms the danger area and establishes near side and far side rally points.

      (d) The PL designates the lead squad to bypass the danger area using the detour bypass method.

      (e) The pace Soldier suspends the current pace count and initiates an interim pace count. The alternate pace and compass Soldier moves forward, offsets the compass 90 degrees left or right as designated, and moves in that direction until clear of the danger area.

      (f) After moving the set distance instructed by the PL, the lead squad assumes original azimuth, and the primary pace Soldier resumes original pace.

      (g) After the open area, the alternate pace and compass Soldier offsets the compass 90 degrees left or right and leads the platoon or squad the same distance (in meters) back to the original azimuth.

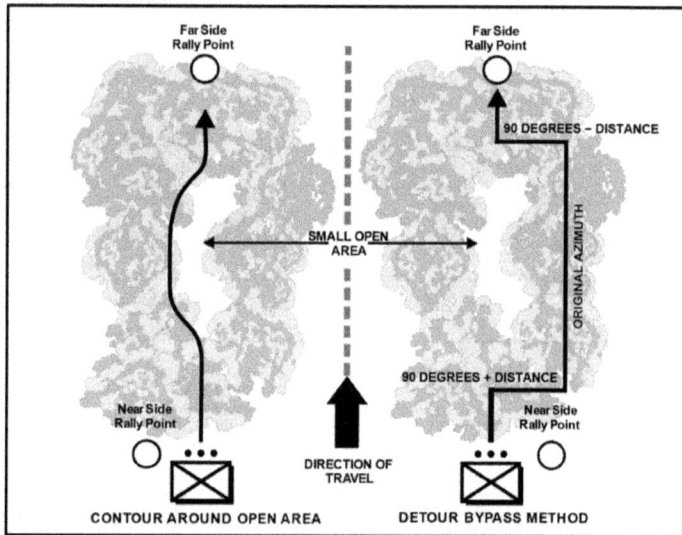

Figure 6-5. Examples of crossing a small, open danger area

(4) <u>Danger areas—series—actions</u>. These are two or more danger areas inside an area that can be either observed or covered by fire.

   (a) Double linear danger area. Units use the linear danger area technique and cross as one.

   (b) Linear or small, open danger area. Units use the bypass or contour technique. (See figure 6-5.)

   (c) Linear or large, open danger area. Units use a platoon wedge formation to cross.

Note. Units cross a series of danger areas with the technique that provides the most security.

(5) <u>Danger area—large—actions</u>.
    (a) The lead squad halts the platoon and signals DANGER AREA.
    (b) The PL moves forward with the RTO and FO to confirm the danger area.
    (c) The PL confirms the danger area and establishes near and far side rally points.
    (d) The PL designates the direction of movement.
    (e) The PL designates any necessary change of formation to ensure security.

Note. The PL plans for all larger danger area crossing sites. Near side security in overwatch sterilizes signs of the patrol. Before a point person steps into a danger area, the PL and FO adjust the targets to cover the movement. When the far side of the danger area is within 250 meters (273 yards), the PL establishes overwatch and designates the lead squad to clear a wood line on the far side.

HAND AND ARM SIGNALS

6-25. Visual signals are any means of communication that require sight and transmit prearranged messages rapidly over short distances. The most common types of visual signals are hand and arm, flag, pyrotechnic, and ground-to-air. Leaders of dismounted units use hand and arm signals to control the movement of individuals, teams, and squads. <u>Units bypass danger areas whenever possible</u>.

6-26. Figures 6-6 through 6-18 on pages 6-14 through 6-17 demonstrate the standardized hand and arm signals for use during patrol operations in Ranger School. Regarding these figures of hand and arm signals—

- Those signals illustrated with a single arrowhead indicate they are not continuously repeated, yet they may be repeated at intervals until acknowledged or the desired action is executed.
- Those signals illustrated with double arrowheads are repeated continuously until acknowledged or the desired action is taken.
- Signals are illustrated as normally seen by the viewer.
- Some signals are illustrated in oblique, right angle, or overhead views for clarity.

Figure 6-6. Assemble or rally signal

Figure 6-7. Join me, follow me, or come forward signal

Figure 6-8. Enemy in sight signal

Figure 6-9. Take cover signal

Figure 6-10. Map check signal

Figure 6-11. Halt signal

Figure 6-12. Take a knee signal

Figure 6-13. Move to the prone signal

Figure 6-14. Pace count signal

Figure 6-15. Headcount signal

Figure 6-16. Danger area signal

Figure 6-17. Freeze signal

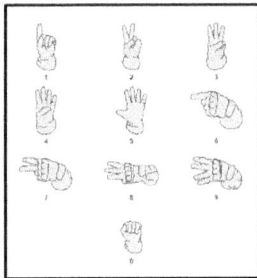

Figure 6-18. Number signals

This page intentionally left blank.

Chapter 7
Patrols

Infantry platoons and squads primarily conduct two types of patrols, reconnaissance and combat. This chapter describes the principles of patrolling, planning considerations, types of patrols, supporting tasks, PBs, and movement to contact. In this chapter, the terms *element* and *team* refer to the squads, fire teams, or buddy teams who perform the tasks as described.

PRINCIPLES

7-1. Five principles govern all patrols: planning, reconnaissance, security, control, and common sense. Further considerations for each of these principles include the following:

- Planning — Quickly make a simple plan and effectively communicate it to the lowest level. A great plan that takes forever to complete and is poorly disseminated is not a great plan. Plan and prepare to a realistic standard and rehearse everything.
- Reconnaissance — A Ranger leader's responsibility is to confirm what they think they know and to learn that which they do not already know.
- Security — Preserve the force as a whole. Every Ranger and every rifle counts; either one has the potential to make the difference between victory and defeat.
- Control — Clarify the concept of operations and commander's intent, coupled with disciplined communications, to bring every Soldier and weapon available to overwhelm the enemy at the decisive point.
- Common sense — Use all available information and good judgment to make sound, timely decisions.

PLANNING

7-2. Planning considerations common to most patrols include task organization, initial planning and coordination, completion of the plan, and contingency planning. A patrol is a detachment sent out by a larger unit to conduct a specific mission.

7-3. Patrols operate semi-independently and return to the main body upon completion of their missions. Patrolling fulfills the Infantry's primary function of finding the enemy to engage them or to report their disposition, location, and actions. Patrols act as the eyes and ears of the larger unit and as a fist to deliver a sharp, devastating jab. Then, they withdraw before the enemy recovers. Some associated definitions with patrols follow:

- Patrol — sent out by a larger unit to conduct a specific combat, reconnaissance, or security mission. A patrol's organization is temporary and specifically matched to the immediate task. Because a patrol is an organization, not a mission, it is not correct to speak of giving a unit the mission to patrol.
- Patrolling (or conducting a patrol) — the semi-independent operation to accomplish the patrol's mission. A patrol requires a specific task and purpose.
- Employment — the commander's sending of a patrol from the main body to conduct a specific tactical task with an associated purpose. Upon completion of that task, the patrol leader returns to the main

body, reports to the commander, and describes the events that took place, the status of the patrol's members and equipment, and any observations that were made.

- Leaders — the SL's responsibility when an organic unit such as a rifle squad composes the patrol; an officer's or noncommissioned officer's designation when mixed elements from several units compose the patrol. This temporary title defines the leader's role and responsibilities for that mission. The patrol leader may designate an assistant—normally the next senior Soldier in the patrol—and any requisite subordinate element leaders.

- Size — a patrol unit is sometimes as small as a fire team. Squad- and platoon-sized patrols are normal. For combat tasks such as a raid, the patrol sometimes comprises most of the combat elements of a rifle company. Unlike operations in which the Infantry platoon or squad is integrated into a larger organization, the patrol is semi-independent and relies on itself for security.

7-4. All patrols and teams share some common elements. Examples of these commonalities follow:

- Reconnaissance and surveillance (R&S) teams are normal for a zone reconnaissance but useful in any situation in which it is impractical to separate the responsibilities of reconnaissance and security. When these responsibilities are separate, the security element provides security at danger areas, secures the objective rally point, isolates the objective, and supports the withdrawal of the rest of the platoon once reconnaissance is complete. The security element sometimes has separate security teams, each with an assigned task or sequence of tasks.

- Assault elements seize and secure the objective and protect special teams as they complete their assigned actions on the objective.

- Security elements provide security at danger areas, secure the objective rally point, isolate the objective, and support the withdrawal of the rest of the patrol once actions on the objective are complete. The security element sometimes has separate security teams, each with an assigned task or sequence of tasks.

- Support elements provide direct and indirect fire support for the unit. Direct fire includes machine guns, medium and light antiarmor weapons, and small recoilless rifles. The available indirect fire includes mortars, artillery, CAS, and organic M320 weapon systems.

- Demolition teams are responsible for preparing and detonating the charges to destroy designated equipment, vehicles, or facilities on the objective.

- Enemy prisoner of war (EPW) and search teams are the assault element who provides two-Ranger (buddy team) or four-Ranger (fire team) search teams to search bunkers, buildings, or tunnels on the objective. These teams search the objective or kill zone for any PIR that provides the PL with an idea of the enemy concept for future operations. Leaders may assign primary and alternate teams to ensure enough prepared personnel are available on the objective.

- Aid and litter teams provide initial treatment and movement of casualties to the platoon's CCP. They are members of the same fire team or squad to ensure proper command and control.

- Breach elements conduct initial penetration of enemy obstacles to seize a foothold and allow the patrol to enter an objective. This action typically accords with the METT-TC (I) variables.

- Reconnaissance teams reconnoiter the objective area from various vantage points once the security teams are in position. Normally, reconnaissance teams are two-Soldier teams to reduce the possibility of detection.

INITIAL PLANNING AND COORDINATION

7-5. Leaders plan and prepare for patrols using the TLP and estimation of the situation as described in chapter 2. Through an estimation of the situation, leaders identify required actions on the objective (mission analysis) and plan backward to departure from friendly lines and forward to reentry of friendly lines. Because patrolling units act independently, they move beyond the direct fire support of the parent unit and operate forward of friendly units. Coordination is thorough, detailed, and continuous throughout planning and preparation. PLs use checklists to avoid omitting any items vital to the accomplishment of the mission.

7-6. Coordination with higher HQ includes intelligence, operations, and fire support. This initial coordination is an integral part of the TLP (step 3—make a tentative plan [see table 2-1 on page 2-1]). The leader also coordinates the unit's patrol activities with the leaders of other units who will be patrolling in adjacent areas at the same time.

COMPLETION OF PLAN

7-7. The PL considers specified and implied tasks while completing the plan. The PL ensures the assigning of all specified tasks to be performed on the objective, at rally points, danger areas, security or surveillance locations, passage lanes, and along routes. These compose the maneuver and the tasks to maneuver units subparagraphs of the execution paragraph.

7-8. The leader also considers key travel and execution times. The leader estimates time requirements for movement to the objective, leader's reconnaissance of the objective, establishment of security and surveillance, completion of all assigned tasks on the objective, and passage through friendly lines. Some planning factors include the following:

- Movement — an average of 1 kilometer per hour (0.6 miles per hour) during daylight hours in woodland terrain; average limited visibility of 0.5 kilometers per hour (0.3 miles per hour). Add time for restrictive or severely restrictive terrain such as mountains, swamps, or thick vegetation.
- Leader's reconnaissance — NLT 1.5 hours.
- Establishment of security and surveillance — one-half hour.

7-9. The leader selects primary and alternate routes to and from the objective. The return routes differ from the routes to the objective. The PL may delegate route selection to a subordinate but is ultimately responsible for the selected routes.

7-10. The leader considers using special signals. Such signals include hand and arm, flare, vocal, whistle, radio, and infrared equipment. Units identify and rehearse primary and alternate signals until all Rangers know their meanings. Leaders consider the following password systems.

- Odd number system, which works as follows:
  - The leader specifies an odd number.
  - The challenge is then any number less than the specified number, and the password is the addend that, when summed with the challenge number, equals the specified number.
  - For example, the specified number is seven, the challenge is three, and the password is four (7 = 3 + 4).
- Running password, which works as follows:
  - This code word alerts a unit to friendly Rangers approaching in a less than organized manner and possibly under pressure.

- The number of Rangers approaching follows the running password.
- For example, the running password is *Ranger* and five friendly Rangers are approaching, so the code word is *Ranger five*.

Note. The plan for maintaining communications normally follows the below order of communications precedence list, the order in which an element moves through available communications systems until it establishes contact with the desired distant element. The Ranger Course standardizes the <u>PACE</u> plan communications systems to the following:
Primary – FM communications
Alternate – Whistle blast (a number combination such as 5 total wherein 3 blasts receive 2 in return)
Contingency – Day – Flaming rock (VS-17 signal panel tied to a rock or similar object)
Night – Infrared chemical light stick on a string of parachute cord
Emergency – Runner

7-11. Always know the location of leaders. The PL considers the locations of the PSG and other key leaders during each phase of the mission. The PL is positioned for best control of the actions of the patrol. The PSG is normally located with the assault element during a raid or attack to help the PL control the use of additional assaulting squads and to assist with securing the objective. The PSG locates at the CCP to facilitate casualty treatment and evacuation. During a reconnaissance mission, the PSG stays behind in the objective rally point to facilitate the transfer of intelligence to the higher HQ and to control the reconnaissance element's movement into and out of the objective rally point.

7-12. A combat patrol collects and reports information gathered during the mission, whether related to the combat task or not. Platoons or squads send SALUTE reports to their controlling HQ when observing enemy activity or making contact with the enemy. It is also important for the leader to communicate timely and accurate situation and status reports including: spot reports; status to the PL (including squad location and progress, enemy situation, enemy KIA, and security posture); status of ammunition, casualties, and equipment to the PSG.

7-13. Unless the mission requires it, the unit avoids enemy contact. The leader's plan addresses actions on enemy contact at each phase of the patrol mission. The unit's ability to continue depends on how early it makes contact, whether the platoon breaks contact without the enemy's detecting its subsequent direction of movement, and whether the unit suffers any casualties from the contact. The plan addresses the handling of seriously wounded Rangers and those KIA. The plan also addresses the handling of prisoners captured due to chance contact and not as part of the planned mission.

7-14. The leader leaves the unit for many reasons throughout the planning, coordination, preparation, and execution of the patrol mission. Each time the leader departs the patrol main body, the leader left in charge of the unit receives a five-point contingency plan. The patrol leader issues additional specific guidance stating the tasks to be accomplished under the leader left in charge. The patrol leader uses the mnemonic device GOTWA to remember and include the going, others, time, what, and action points of the contingency plan (see table 7-1).

Table 7-1. Going–others–time–what–action format

Acronym Letters	Points the Plan Addresses	Questions the Plan Answers
G	Going	Where is the leader going?
O	Others	Who is going with the leader?
T	Time (Duration)	How long will the element be gone? Or when will they return?
W	What	What should happen if the leader were to fail to return?
A	Action	What actions will the element and the main body execute upon enemy contact?

7-15 The leader considers the use and location of rally points. A rally point is a place the leader designates for a displaced unit to reassemble and reorganize. Rangers know to which rally point to move at each phase of the patrol mission upon their separation from the unit. They also know the requisite actions there and how long they are to wait at each rally point before moving to another. The most common types of rally points include initial, en route, objective, near side, and far side. Rally points possess the following characteristics:

- Easily identifiable in daylight and limited visibility.
- Show no signs of recent enemy activity.
- Covered and concealed from the ground and air.
- Away from natural lines of drift and high-speed AAs.
- Defensible for short periods of time.

7-16 The objective rally point typically lies 200 to 400 meters (219 to 437 yards) from the objective or, at a minimum, one major terrain feature away. Actions at the objective rally point include the following:

- Stopping to look, listen, and smell before pinpointing location.
- Conducting a leader's reconnaissance of the objective.
- Issuing any necessary FRAGORDs.
- Making final preparations before continuing operations (for example, reapply camouflage; prepare demolitions; line up rucksacks for quick recovery; prepare EPW's bindings, first aid kits, and litters; inspect weapons).
- Accounting for Rangers and equipment after completing actions on the objective.
- Disseminating information from reconnaissance when no contact was made.

7-17 The plan includes a leader's reconnaissance of the objective once the platoon or squad establishes the objective rally point. Before departing, the leader issues a five-point contingency plan. During reconnaissance, the leader pinpoints the objective; selects reconnaissance, security, support, and assault positions for the elements; and adjusts the plan based on observation of the objective.

7-18 Each type of patrol requires different tasks during the leader's reconnaissance. The leader brings along different elements. These are discussed separately under each type of patrol. The leader plans time to return to the objective rally point, complete the plan, disseminate information, issue orders and instructions, and allow the squads to make any additional preparations. During the leader's reconnaissance for a raid or ambush, the PL leaves surveillance on the objective. Each type of patrol also requires different actions on the objective. Actions on the objective are discussed under each type of patrol.

RECONNAISSANCE PATROLS

7-19. Area and zone reconnaissance patrols provide timely and accurate information on the enemy and terrain and confirm the leader's plan before its execution. Units on reconnaissance operations collect specific information (PIR) or general information based on the instructions from their higher commanders.

7-20. For successful area reconnaissance, the PL applies the fundamentals of reconnaissance to the plan while conducting the operation. To obtain required information, the parent unit tells the patrol leader the necessary information in the form of the information requirement and PIRs. The platoon's mission is then tailored to the required information. During the entire patrol, members continuously gain and exchange all gathered information but do not consider the mission accomplished until all the PIRs have been gathered.

7-21. A patrol avoids letting the enemy know it is in the objective area. When the enemy knows they are under observation, they tend to move, change plans, or increase security measures. Methods of avoiding the enemy's detection follow:

- Minimizing movement in the objective area (area reconnaissance).
- Moving no closer to the enemy than necessary.
- Using long-range surveillance or NVDs whenever possible.
- Using camouflage, stealth, and noise and light disciplines.
- Minimizing radio traffic.

7-22. A patrol retains an ability to break contact and return to the friendly unit with the gathered information. When necessary, they break contact and continue the mission. Leaders emplace security elements to overwatch the reconnaissance elements. They suppress the enemy to enable the reconnaissance element to break contact.

7-23. When the PL receives the order, the mission undergoes analysis to ensure understanding of what is to be done. The PL task-organizes the platoon to accomplish the mission in accordance with the METT-TC (I) variables. A reconnaissance mission is typically a squad-sized mission. The types of reconnaissance are—

- Area reconnaissance — The area reconnaissance patrol collects all available information for the area on the PIR and other intelligence the order does not specify. It completes reconnaissance and reports all information by the time the order specifies. The patrol is not compromised.
- Zone reconnaissance — The zone reconnaissance patrol determines for its assigned zone all the PIRs and other intelligence the order does not specify. It reconnoiters without the enemy's detection. It completes reconnaissance and reports all information by the time the order specifies.

ACTIONS ON THE OBJECTIVE, AREA RECONNAISSANCE

7-24. The element occupies the objective rally point as discussed in paragraph 7-47. The RTO reports to higher HQ that the unit has occupied the objective rally point. The leader confirms the location on the map while subordinate leaders make necessary perimeter adjustments. (See figure 7-1.) The PL organizes the platoon in one of two ways: separate reconnaissance and security elements or combined reconnaissance and security elements.

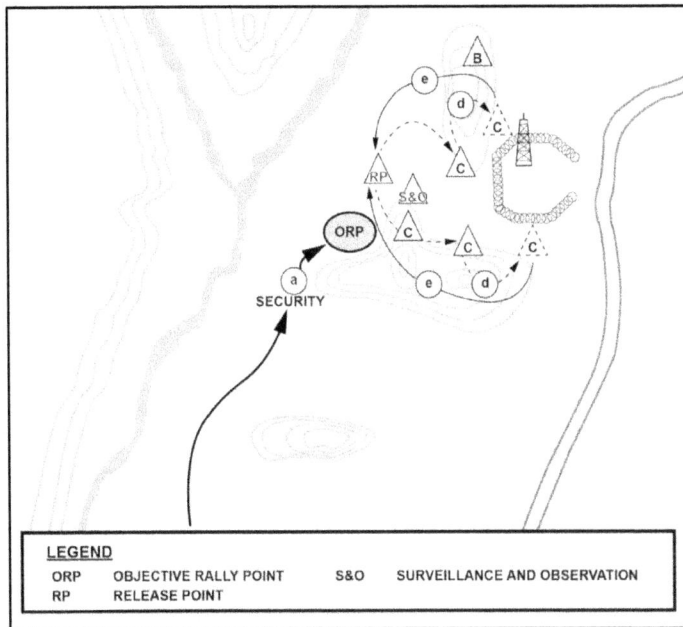

LEGEND

ORP	OBJECTIVE RALLY POINT	S&O	SURVEILLANCE AND OBSERVATION
RP	RELEASE POINT		

Figure 7-1. Example of area reconnaissance

7-25. The PL takes subordinate leaders and key personnel on a leader's reconnaissance to confirm the objective and plan. The PL also—

- Issues a five-point contingency plan before departure.
- Establishes a suitable RP that is definitely out of sight but, whenever possible, is also beyond sight and sound of the objective. In addition, the suitable RP has good rally point characteristics.

- Allows all personnel to become familiar with the RP and surrounding area.
- Identifies the objective and emplaces surveillance, designates a surveillance team to keep the objective under surveillance, and issues a contingency plan to the senior Soldier remaining with the surveillance team. The surveillance team is positioned with one Soldier facing the objective and one facing back in the direction of the RP.
- Takes subordinate leaders forward to pinpoint the objective, establish a limit of advance (LOA), and choose vantage points.
- Maintains communications with the platoon throughout the leader's reconnaissance.

7-26. The PSG maintains security, supervises the priorities of work in the objective rally point, and reestablishes security at the objective rally point. The PSG also disseminates the PL's contingency plan and oversees the preparation of the reconnaissance personnel (for example, Soldiers have camouflage reapplied; NVDs and binoculars prepared; weapons on safe with a round in the chamber).

7-27. The PL and reconnaissance party return to the objective rally point. The PL confirms the plan or issues a FRAGORD and allows subordinate leaders time to disseminate the plan.

7-28. The patrol conducts reconnaissance by long-range observation and surveillance whenever possible. The R&S element moves to observation points offering cover and concealment and outside small arms' range, establishes a series of OPs when one location does not suffice for gathering information, and gathers all the PIRs using the SALUTE format.

7-29. If necessary, the patrol conducts reconnaissance by short-range observation and surveillance (see figure 7-1 on page 7-7), moves to an OP near the objective, passes close enough to the objective to gain information, and gathers all the PIRs using the SALUTE format.

7-30. R&S teams use techniques such as the cloverleaf method to move to successive OPs. (See figure 7-1 on page 7-7.) In this method, R&S teams avoid paralleling the objective site, maintain extreme stealth, do not cross the LOA, and maximize the use of available cover and concealment.

7-31. During the conduct of reconnaissance, each R&S team returns to the RP when any of the following occurs:
- They have gathered all their PIRs.
- They have reached the LOA.
- The allocated time to conduct reconnaissance has elapsed.
- They make enemy contact.

7-32. At the RP, the leader analyzes the gathered information and determines whether it meets the PIRs. When the leader determines the gathered information insufficient to meet the PIRs, or when the information differs drastically from that which the leader and subordinate leader gathered, the leader may send the R&S teams back to the objective site. In this case, the R&S teams alternate areas of responsibilities. For example, when one team previously reconnoitered from the 6–3–12, that team now reconnoiters from the 6–9–12.

7-33. The R&S element returns undetected to the objective rally point by the specified time, disseminates information to all patrol members through key leaders, or moves to a position at least one terrain feature or 1 kilometer (0.6 miles) away to disseminate. To disseminate, the leader has the RTO prepare three sketches of the objective site based on the leader's sketch and provides copies to the subordinate leaders to assist in dissemination. The R&S element reports any information requirements or any information requiring immediate attention to higher HQ and departs for the designated area.

7-34. If the R&S element makes contact, it moves to the RP. The reconnaissance element tries to break contact and return to the objective rally point. There, it secures rucksacks and quickly moves out of the area. Once it has moved a safe distance away, the leader informs higher HQ of the situation and takes further instructions from them.

    a. While emplacing surveillance, the reconnaissance element withdraws through the RP to the objective rally point and follows the same procedures as in paragraph 7-32.

    b. While conducting reconnaissance, the compromised element returns a sufficient volume of fire to break contact. Surveillance fires an AT4 (antitank) at the largest weapon on the objective. All elements pull off the objective and move to the RP. The senior leader quickly accounts for all personnel and returns to the objective rally point. Once in the objective rally point, leaders follow the procedures that paragraph 7-32 describes. The critical tasks for a patrol follow.

        (1) Secure and occupy the objective rally point.

        (2) Conduct a leader's reconnaissance of the objective.

           (a) Estimate the RP.

           (b) Pinpoint the objective.

           (c) Emplace surveillance via the surveillance and observation team. Position any security element.

        (3) Conduct reconnaissance by long-range surveillance whenever possible.

        (4) Conduct reconnaissance by short-range surveillance as necessary.

        (5) Teams.

           (a) Move to successive OPs when necessary.

           (b) On order, return to the RP.

           (c) Once the gathering of the PIR is complete, return to the objective rally point.

        (6) Patrol.

           (a) Link up as directed in the objective rally point.

           (b) Disseminate information before moving.

ACTIONS ON THE OBJECTIVE, ZONE RECONNAISSANCE

7-35. The element occupies the initial objective rally point as discussed in paragraph 7-47. The radio operator calls in SPARE for the occupation of the objective rally point. The leader confirms the location on the map while subordinate leaders make necessary perimeter adjustments.

    a. <u>Organization</u>. The reconnaissance TLs organize their reconnaissance elements.

        (1) Designate security and reconnaissance elements.

        (2) Assign responsibilities such as point person, pace person, en route recorder, and rear security when not already assigned.

        (3) Designate easily recognizable rally points.

        (4) Ensure local security at all halts.

    b. <u>Actions</u>. The patrol reconnoiters the zone.

        (1) Moves tactically to the objective rally points.

        (2) Occupies the designated objective rally points.

        (3) Follows the method designated by the PL, such as fan, converging routes, or box. (See table 7-2 on page 7-10.)

(4) The reconnaissance teams reconnoiter.
  (a) During movement, the squad gathers all the PIRs that the order specifies.
  (b) The reconnaissance TLs ensure their teams draw sketches or take digital photos of all the enemy hard sites, roads, and trails.
  (c) Return to the objective rally point or link up at the rendezvous point on time.
  (d) When the squad arrives at a new rendezvous point or objective rally point, the reconnaissance TLs report all the gathered information to the PL.
(5) The PL continues to control the reconnaissance elements.
  (a) Moves with the reconnaissance element who establishes the rendezvous point.
  (b) Changes reconnaissance methods as required.
  (c) Designates times for the elements to return to the objective rally point or to link up.
  (d) Collects all information and disseminates it to the entire patrol. The PL briefs all key subordinate leaders on information gathered by other squads—establishing one consolidated sketch whenever possible—and allows TLs time to brief their teams.
  (e) With the PSG, accounts for all personnel.
(6) The patrol continues reconnaissance until it has reconnoitered all the designated areas and then returns undetected to friendly lines.

Table 7-2. Reconnaissance methods

FAN METHOD	Use a series (fan) of ORPs. Patrol establishes security at the first ORP. Each reconnaissance element moves from the ORP along a different fan-shaped route. Each route overlaps with those of other reconnaissance elements. This ensures reconnaissance of the entire area. The leader maintains a reserve at the ORP. Once all the reconnaissance elements return to the ORP, the PL collects and disseminates all information before moving to the next ORP.
CONVERGING ROUTES METHOD	The PL selects routes from the ORP through the zone to a rendezvous point at the far side of the zone from the ORP. Each reconnaissance element moves and reconnoiters along a specified route. They converge, or link up, at one time and place.
BOX METHOD	The PL sends reconnaissance elements from the first ORP along routes that form a box. Then, the PL sends other elements along routes throughout the box. All teams link up at the far side of the box from the ORP.
Legend: ORP–objective rally point; PL–platoon leader	

COMBAT PATROLS

7-36. Combat patrols are the second type of patrol. Combat patrols receive further division into raids, ambushes, and security patrols. Units conduct combat patrols to destroy or capture enemy soldiers or equipment; destroy installations, facilities, or key points; or harass enemy forces. Combat patrols also provide security for larger units.

7-37. In planning a combat patrol, the PL considers the following:

a. Tasks to maneuver units. Normally the platoon HQ element controls the patrol on a combat patrol mission. The PL makes every effort to maintain squad and fire team integrity as subordinate units receive their assigned tasks.

    (1) The PL considers the requirements for assaulting the objective and supporting the assault by fire and the security of the entire unit throughout the mission.

        (a) For the assault on the objective, the PL considers the required actions on the objective, the size of the objective, and the known or presumed strength and disposition of the enemy on and near the objective.

        (b) The PL considers the available weapons and the type and volume of fire required to provide fire support for the assault on the objective.

        (c) The PL considers the requirement to secure the platoon at points along the route, at danger areas, at the objective rally point, along enemy AAs into the objective, and elsewhere during the mission.

        (d) The PL also designates engagement and disengagement criteria.

    (2) The PL assigns additional tasks to the squads for demolition, EPW searches, guarding EPWs, treating and evacuating (by litter teams) friendly casualties, and other requirements for the successful completion of the patrol mission when the SOP does not.

    (3) The PL determines who controls any attachments of skilled personnel or special equipment.

b. Leader's reconnaissance of the objective. In a combat patrol, the PL has additional considerations for the conduct of reconnaissance of the objective from the objective rally point.

    (1) Composition of the leader's reconnaissance party. The PL normally brings the following personnel:

        (a) SLs including the WSL.

        (b) Surveillance team.

        (c) FO.

        (d) Security element depending on the time available.

    (2) Conduct of the leader's reconnaissance. In a combat patrol, the PL considers the following additional actions in the conduct of the leader's reconnaissance of the objective.

        (a) The PL designates an RP about halfway between the objective rally point and the objective, using the same characteristics as a rally point. The PL then issues a five-point contingency plan to the security element, and the PL, FO, and surveillance team move to pinpoint the objective and emplace the surveillance team with eyes on the objective. The PL next moves back to the RP and emplaces the security element.

        (b) The PL confirms the location of the objective or kill zone, notes the terrain, and identifies the emplacements of claymore mines to cover dead space. SLs receive any changes to the plan while overlooking the objective whenever possible.

        (c) When the objective is the kill zone for an ambush, the leader's reconnaissance party does not cross the objective, to do so leaves tracks that can compromise the mission.

        (d) The PL confirms the suitability of the assault and support positions and routes from them back to the objective rally point.

        (e) The PL issues a five-point contingency plan to the surveillance team before returning to the objective rally point.

AMBUSH

7-38. An ambush is a surprise attack from a concealed position on a moving or temporarily halted target. It falls into the hasty or deliberate category and point or area type. It assumes a linear or L-shaped formation.

7-39. The leader considers various key factors to determine the ambush category, type, and formation. Then, the leader develops the ambush plan and considers the following while planning and executing an ambush:

- Key factors:
  - Coverage (of ideally the whole kill zone) by fire.
  - METT-TC (I) variables.
  - Existing or reinforcing obstacles including claymore mines to keep the enemy in the kill zone.
  - Security teams usually with hand-held antitank weapons such as AT4s or light antitank weapons, claymore mines, and various means of communication.
  - Security elements or teams to isolate the kill zone.
  - Claymore mines or explosives for protecting the assault and support elements.
  - Proper method of movement through the kill zone to the LOA that allows some portion of the assaulting element to provide overwatch from a stable firing position.

Note. The assault element retains an ability to move quickly through its own protective obstacles.

  - Timing of all the platoon elements' actions to preclude loss of surprise.

Note. In the event of the compromise of any member of the ambush, the leader may immediately initiate the ambush.

When employing an ambush of long duration, one squad conducts the entire ambush. Determine the movement time of the rotating squads from the objective rally point to the ambush site.

- Categories:
  - Hasty — A unit conducts a hasty ambush when it makes visual contact with an enemy force and has time to establish an ambush without the enemy's detection. The actions for a hasty ambush are well rehearsed so Rangers know what to do on the leader's signal. They also know what actions to take when the enemy detects the unit before it is ready to initiate the ambush.
  - Deliberate — A deliberate ambush is conducted at a predetermined location against any enemy element who meets the commander's engagement criteria. The leader requires the following detailed information when planning a deliberate ambush: the size and composition of the targeted enemy and the weapons and equipment available to the enemy.
- Types:
  - Point — Rangers deploy to attack an enemy in a single kill zone.
  - Area — Rangers deploy in two or more related point ambushes.
  - Antiarmor — Rangers focus on moving or temporarily halted enemy armored vehicles.

- Formations (see figure 7-2):
  - Linear — The assault and support elements deploy parallel to the enemy's route. This positions both elements on the long axis of the kill zone and subjects the enemy to flanking fire. This formation is useful in close terrain that restricts the enemy's ability to maneuver against the platoon and also in open terrain, provided a means of keeping the enemy in the kill zone is effectible
  - L-shaped — The assault element forms the long leg parallel to the enemy's direction of movement along the kill zone. The support element forms the short leg at one end and at a right angle to the assault element. This provides both flanking (long leg) and enfilading (short leg) fires against the enemy. This formation is useful at a sharp bend in a trail, road, or stream but not where the short leg crosses a straight road or trail

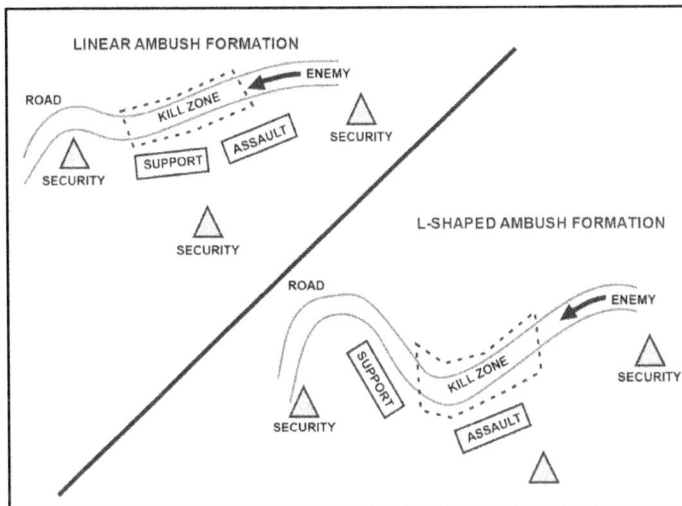

Figure 7-2. Ambush formations

HASTY AMBUSH

7-40. The platoon moves quickly to concealed positions. The ambush initiation occurs only after the majority of the enemy is in the kill zone. The unit does not become decisively engaged but surprises the enemy. The patrol captures, kills, or forces the withdrawal of the entire enemy inside the kill zone.

7-41. On order, the patrol withdraws all personnel and equipment in the kill zone from observation and direct fire. The unit does not become decisively engaged by follow-on elements. The platoon continues follow-on operations. Actions on the objective follow (see figure 7-3).

   a. Using visual signals, any Ranger alerts the unit that an enemy force is in sight. The Ranger continues to monitor the location and activities of the enemy force until the TL or SL relieves the Ranger and gives the enemy location and direction of movement.

   b. The platoon or squad halts and remains motionless.

      (1) The PL gives the signal to conduct a hasty ambush, taking care not to alert the enemy of the patrol's presence.

      (2) The leader determines the best nearby location for a hasty ambush and uses hand and arm signals to direct the unit members to covered and concealed positions.

   c. The leader designates the location and extent of the kill zone.

   d. The teams and squads move silently to covered and concealed positions, ensuring positions avoid the enemy's detection and have good observation and fields of fire into the kill zone.

   e. The security elements move out to cover each flank and the rear of the unit. The leader directs the security elements to move a given distance, set up, and then rejoin the unit on order or after the ambush (when the sound of firing ceases). At squad level, the two outside buddy teams normally provide flank security, as well as fires into the kill zone. At platoon level, fire teams compose the security elements.

   f. The PL assigns sectors of fire and issues any other necessary commands such as control measures.

   g. The PL initiates the ambush—using the greatest casualty-producing weapon available—when the largest percentage of the enemy is in the kill zone.

   h. The PL controls the rate and distribution of fire, employs indirect fire to support the ambush, and orders cease-fire. When the situation dictates, the PL orders the patrol to assault through the kill zone.

   i. The PL designates personnel to conduct a hasty search of enemy personnel and then to process EPWs and their equipment.

   j. The PL orders the platoon to withdraw from the ambush site along a covered and concealed route.

   k. The PL gains accountability, appropriately reorganizes, disseminates information, reports the situation, and continues the mission as directed.

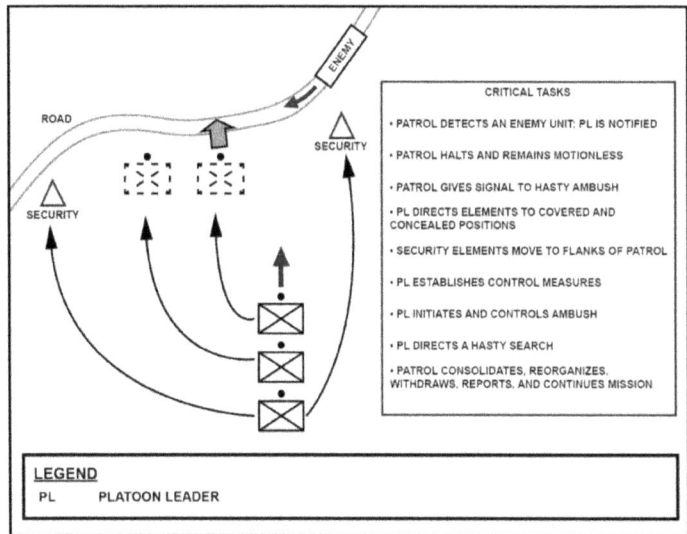

Figure 7-3. Hasty ambush

DELIBERATE (POINT/AREA) AMBUSH

7-42. The ambush emplacement occurs NLT the time the order specifies. The patrol surprises the enemy and engages their main body, killing or capturing all enemy in the kill zone and destroying equipment within the commander's intent. The patrol withdraws all personnel and equipment from the objective, on order, within the time the order specifies. The patrol obtains all available PIRs from the ambush and continues follow-on operations. Actions on the objective follow (see figure 7-4 on page 7-16).

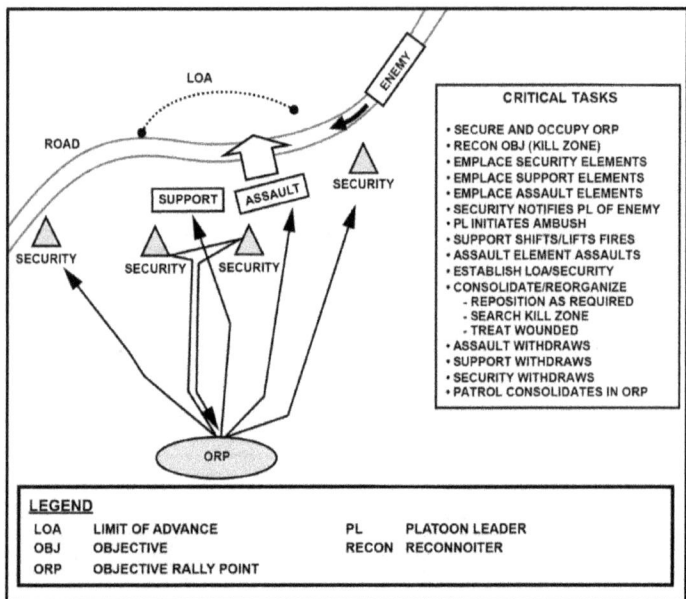

Figure 7-4. Deliberate ambush

a. The PL prepares the patrol for the ambush in the objective rally point.
b. The PL prepares to conduct a leader's reconnaissance.
   (1) Designates the members of the leader's reconnaissance party such as SLs, a surveillance team, an FO, and possibly the security element.
   (2) Issues a contingency plan to the PSG.
c. The PL conducts the leader's reconnaissance.
   (1) Ensures the leader's reconnaissance party moves without the enemy's detection.

(2) Confirms the objective location and suitability for the ambush.
(3) Selects a kill zone.
(4) Posts the surveillance team at the site and issues a contingency plan.

Note. Security teams occupy prior to the PL reconnoitering the assault or support by fire positions, securing the flanks of the ambush site and providing early warning. A security team remains in the objective rally point whenever the patrol plans to return to the objective rally point after actions on the objective. Upon abandonment of the objective rally point, the emplacement of a rear security team occurs after PLAN and before CONFIRMS. The objective is secured prior to initiating reconnaissance of the assault or support by fire positions.

---

   (5) Confirms suitability of assault and support positions and routes from the positions to the objective rally point.
   (6) Selects positions for each weapon system in support by fire positions and then designates sectors of fire.
   (7) Identifies all offensive control measures to be used. Identifies the probable line of deployment, assault position, LOA, and any boundaries or other control measures. The PL may use any available infrared aiming devices to identify these positions on the ground.
 d. The PL adjusts the plan based on information from reconnaissance.
   (1) Assigns positions.
   (2) Designates withdrawal routes.
 e. The PL confirms the ambush formation.
 f. The support element leader assigns sectors of fire.
   (1) Emplaces claymore mines and obstacles as designated.
   (2) Identifies sectors of fire and emplaces limiting stakes or uses metal-to-metal contact with the machine gun tripod to prevent fratricide on the objective.
   (3) Overwatches the movement of the assault element into position.
 g. Once the support element is in position or on the PL's order, the assault element departs the objective rally point and moves into position.
 h. Upon reaching the assault position, leaders identify individual sectors of fire as assigned by the PL. The support element emplaces aiming stakes and, to help destroy the enemy in the kill zone, claymore mines and camouflages positions.
 i. The security element spots the enemy and notifies the PL with reports on the direction of movement, size of the target, and any special weapons or equipment. The security element also keeps the PL informed of any enemy forces following the lead force.
 j. The PL alerts other elements and determines whether the enemy force is too large for the ambush to engage successfully.
 k. The PL initiates the ambush with the highest casualty-producing device. The PL may use a command-detonated claymore mine and plan a backup method for initiating the ambush in case the primary means fail. In the event the PL does not use a claymore mine or the mine misfires, the PL initiates with a closed bolt weapon system. Open bolt weapon systems are more prone to malfunctions,

thereby providing the enemy with both early warning and the ability to react. All Rangers receive this information and practice it during rehearsals.

l. The PL ensures the assault and support elements deliver fires on the enemy in the kill zone with the heaviest, most accurate volume possible. In limited visibility, the PL may use infrared lasers to define specific targets in the kill zone.

m. Before assaulting the target, the PL gives the signal to lift or shift fires and receives confirmation.

n. The assault element assaults before the remaining enemy reacts, kills or captures the enemy in the kill zone, and uses individual movement techniques or bounds by fire teams to move.

o. Upon reaching the LOA, the assault element halts and establishes security. The element reestablishes the chain of command and key weapon systems as necessary. All Rangers load a fresh magazine or drum of ammunition using the buddy system. LACE report submissions pass through the chain of command. The PL submits an initial contact report to higher HQ.

p. The PL directs special teams (for example, EPW search, aid and litter, demolition) to accomplish their assigned tasks once the assault element has established its LOA.

(1) Once the kill zone is clear, collect and secure all EPWs and move them out of the kill zone before searching their bodies. Coordinate for an EPW exchange point to link up with higher HQ to extract all EPWs. Treat them in accordance with the 5-S rule.

(2) Search from one side to the other and mark the searched bodies to ensure the area is thoroughly covered. Use clear out / search in the intelligence technique, which has the unit clear from the center of the objective outward, to ensure the area is clear of all enemy combatants. At this time, the platoon begins simultaneously conducting any necessary casualty care operations. Then, search all enemy personnel toward the center of the objective. Search all dead enemy personnel using the two-Ranger search technique.

(a) As the search team approaches a dead enemy soldier, one Ranger guards while the other Ranger searches. First, kick the enemy's weapon away.

(b) Search Rangers roll over any prone-positioned body (facedown) by lying on top and, when given the go-ahead by the guard positioned at the enemy's head, rolling the body over onto their own bodies. This protects a search Ranger in case the enemy soldier is concealing a grenade with the pin pulled.

(c) Search Rangers conduct a systematic search of the dead soldier from head to toe, removing all papers and anything new (for example, different type rank, shoulder boards, different unit patch, pistol, weapon, NVD). They note whether the enemy soldier has a fresh or shabby haircut, the condition of the uniform and boots, and the radio frequency. They secure the signal operating instructions, maps, documents, and overlays.

(d) The two Rangers thoroughly search one body at a time and continue in this manner until their platoon has examined all enemy personnel in and near the kill zone.

(3) Identify, collect, and prepare all equipment to be carried back or destroyed.

(4) Evacuate and treat friendly wounded first, then enemy wounded as time permits.

(5) The demolition team prepares dual-primed explosives or incendiary grenades and awaits the signal to initiate. This is normally the last action performed before the unit departs the objective and signals the security elements to return to the objective rally point.

(6) All actions on the objective with a stationary assault line are the same, with the exception of the search teams. To provide security within the teams to the far side of the kill zone during the search,

they work in three-Ranger teams. Before the search begins, the Rangers move all KIAs to the near side of the kill zone.

q. If enemy reinforcements try to penetrate the kill zone, the flank security engages to prevent the compromise of the assault element.

r. The PL directs the units' withdrawal from the ambush site.

(1) Elements normally withdraw in the reverse order from that they followed to establish their positions.

(2) The elements return to the RP or directly to the objective rally point depending on the distance between elements.

(3) The security element of the objective rally point is alert to assist the platoon's return. It maintains security for the objective rally point while the rest of the platoon prepares to leave.

(4) When possible, all elements return to the location at which they separated from the main body. This location is normally the RP.

s. The PL and PSG direct actions at the objective rally point including accountability of personnel and equipment and recovery of rucksacks and other equipment left at the objective rally point during the ambush.

t. The PL disseminates information or first moves the platoon to a safe location (no nearer 1 kilometer [0.6 miles] or one terrain feature away from the objective) and then disseminates information.

u. As required, the PL and FO execute indirect fire to cover the platoon's withdrawal.

PERFORMING A RAID

7-43. Raids are surprise attacks against a position or installation for a specific purpose other than seizing and holding the terrain. Their purpose is to destroy a position or installation, to destroy or capture enemy soldiers or equipment, or to free prisoners. A raid patrol retains terrain just long enough to accomplish the intent of the raid. A raid always ends with a planned withdrawal off the objective and a return to the main body.

7-44. The patrol initiates the raid NLT the time the order specifies, surprises the enemy, assaults the objective, and accomplishes its assigned mission within the commander's intent. The patrol does not become decisively engaged en route to the objective. The patrol obtains all available PIRs from the raid objective and continues follow-on operations.

a. Planning considerations. A raid is a form of attack, usually small in scale, involving a swift entry into hostile territory to secure information, confuse the enemy, or destroy installations and followed by a planned withdrawal. Squads do not conduct raids. The sequence of platoon actions for a raid is similar to that for an ambush. Additionally, the assault element of the platoon conducts any necessary breach of an obstacle and performs any additional tasks on the objective, such as demolition of fixed facilities. The fundamentals of a raid follow.

(1) Surprise and speed. Infiltrate and surprise the enemy without their detection.

(2) Coordinated fires. Seal off the objective with well-synchronized direct and indirect fires.

(3) Violence of action. Overwhelm the enemy with fire and maneuver.

(4) Planned withdrawal. Withdraw from the objective in an organized manner, maintaining security and accountability.

b. Actions on the objective (raid) (see figure 7-5 on page 7-20).

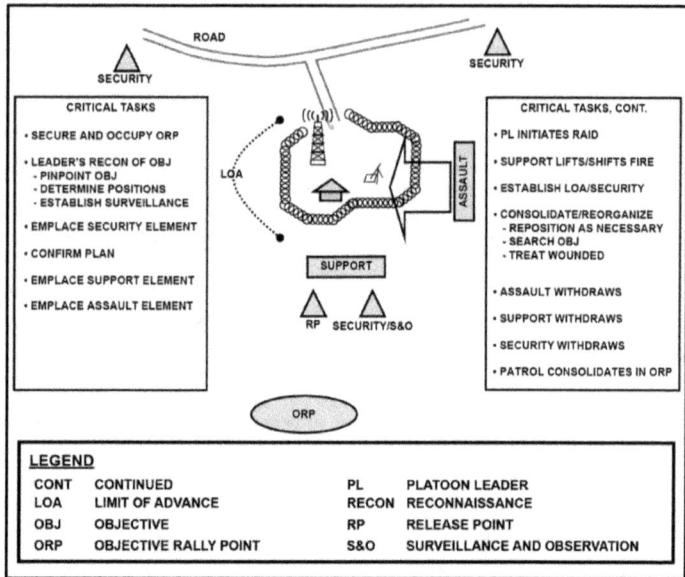

Figure 7-5. Actions on the objective (raid)

(1) The patrol moves to and occupies the objective rally point in accordance with the patrol's SOP. The patrol prepares for the leader's reconnaissance.

(2) The PL, SLs, and selected personnel conduct a leader's reconnaissance.

    (a) The PL establishes the RP; pinpoints the objective; contacts the PSG to prepare Soldiers, weapons, and equipment; emplaces the surveillance team to observe the objective; and verifies and updates intelligence information.

    (b) Upon emplacing the surveillance team, the PL provides a five-point contingency plan to the PSG.

    (c) The security teams are brought forward on the leader's reconnaissance and emplaced before the leader's reconnaissance leaves the RP.

    (d) The leader's reconnaissance verifies the locations of and routes to security, support, and assault positions.

    (e) The leaders conduct reconnaissance without compromising the patrol.

    (f) The leaders normally conduct reconnaissance of the support by fire position first, then the assault position.

(3) The PL confirms, denies, or modifies the plan and issues instructions to the SLs.

    (a) Assigns positions and withdrawal routes to all elements.

    (b) Designates control measures on the objective, such as element objectives, lanes, LOAs, target reference points, and assault lines.

    (c) Affords the SLs time to disseminate information and confirm their elements are ready.

(4) The security elements occupy designated positions, moving without the enemy's detection into positions that provide early warning and can seal off the objective from outside support or reinforcement.

(5) The support element leader moves the support element to their designated positions. The support element leader ensures the element can place well-aimed fire on the objective.

(6) The PL moves with the assault element into the assault position. The assault position is normally the last covered and concealed position before reaching the objective. As it passes through the assault position, the platoon deploys into its assault formation. Its squads and fire teams deploy to place the bulk of their firepower to the front as they assault the objective.

(7) The squad and fire teams also make contact with the surveillance team to confirm any enemy activity on the objective; ensure the assault position is close enough for immediate assault in case the element suffers early detection; move into position without the enemy's detection; and establish local security and fire control measures.

(8) The element leaders inform the PL when their elements are in position and ready.

(9) The PL initiates the raid and directs the support element to fire.

(10) Upon gaining fire superiority, the PL directs the assault element to move toward the objective:

    (a) The assault element holds fire until engaged or until ready to penetrate the objective.

    (b) The PL signals the support element to lift or shift fires and receives confirmation from the support element. The element lifts or shifts fires as directed, shifting fires to the flanks of targets or areas as the order directs.

(11) The assault element attacks and secures the objective. The assault element breaches any wire obstacles. As the platoon or its assault element moves onto the objective, it increases the volume and accuracy of fire. The SLs assign specific targets or objectives for their fire teams. Only when direct fire maintains enemy suppression does the rest of the unit maneuver. As the assault element gets closer to the enemy, emphasis on suppression increases and maneuver decreases. Ultimately, all but one fire team suppresses to allow that one fire team to break into the enemy position. Throughout the assault, Rangers use proper individual movement techniques, and fire teams retain their basic shallow wedge formation. The platoon does not get on line to sweep across the objective.

    (a) The assault element assaults through the objective to the designated LOA.

    (b) The assault element leaders establish local security along the LOA and consolidate and reorganize appropriately. They provide LACE reports to the PL and PSG. The platoon establishes security, reorganizes key weapons, provides first aid, and prepares wounded Rangers for MEDEVAC. They redistribute ammunition and supplies and relocate selected weapons to alternate positions when leaders believe the enemy has pinpointed them during the attack. They adjust other positions for mutual support. The SL and TL provide LACE reports to the PL. The PL or PSG reorganizes the patrol based on the contact.

   _1_ On order, the special teams accomplish all assigned tasks under the supervision of the PL, who is positioned to control the patrol.

   _2_ The special TLs report to PL when their teams complete their assigned tasks.

(12) On order or the signal of the PL, the assault element withdraws from the objective. Using prearranged signals, the assault line begins an organized withdrawal from the objective site, maintaining control and security throughout the withdrawal. The assault element bounds back near the original assault line and begins a single file withdrawal through the antipersonnel landmine chokepoint. All the Rangers move through the chokepoint for an accurate count. Once the assault element is a safe distance from the objective and the headcount is confirmed, the platoon withdraws the support element. When the support elements were a part of the assault line, they withdraw together, and security receives the signal to withdraw. Once the support is a safe distance off the objective, they notify the PL, who contacts the security element and signals them to withdraw. All security teams link up at the RP and notify the PL before moving to the objective rally point. Personnel returning to the objective rally point immediately secure their equipment and establish all-around security. Once the security element returns, the platoon moves out of the objective area as soon as possible.

  (a) Before withdrawing, the demolition team activates devices and charges.

  (b) The support element or designated personnel in the assault element maintain local security during the withdrawal.

  (c) The leaders report updated accountability and status to the PL and PSG.

(13) The squads withdraw from the objective to the objective rally point in the order that the order designates.

  (a) Account for personnel and equipment.

  (b) Disseminate information.

  (c) Redistribute ammunition and equipment appropriately.

(14) The PL reports mission accomplishment to higher HQ and continues the mission.

  (a) Reports raid assessment to higher HQ.

  (b) Informs higher HQ of any information requirements and PIRs gathered.

SUPPORTING TASKS

7-45. Supporting tasks include a linkup, patrol debriefing, and occupation of an objective rally point. A linkup is a meeting of friendly ground forces. It depends on control, detailed planning, communications, and stealth.

  a. Task standard, or the unit's linkup at the time and place the order specifies. The enemy does not surprise the main bodies. The linkup units establish a consolidated chain of command.

  b. Site selection. The leader identifies a tentative linkup site by map reconnaissance or other imagery, or higher HQ designates a linkup site. The best linkup site has the following characteristics.

    (1) Ease of recognition.

    (2) Cover and concealment from ground and air.

(3) No tactical value to the enemy.
(4) Location away from natural lines of drift.
(5) Defensible for a short period of time.
(6) Multiple access and escape routes.

c. Execution. The linkup procedure begins as the unit moves to the linkup point. The steps of this procedure follow.

(1) The stationary unit performs linkup actions.
  (a) Occupies the linkup rally point NLT the time the order specifies.
  (b) Establishes all-around security and communications and prepares to accept the moving unit.
  (c) The security team clears the immediate area around the linkup point; marks the linkup point with the coordinated recognition signal; moves to a covered and concealed position; and observes the linkup point and immediate area around it.

(2) The moving unit.
  (a) Performs linkup actions.
  (b) Reports its location using phase lines, checkpoints, or other control measures.
  (c) Halts at a safe distance from the linkup point in a covered and concealed position (the linkup rally point).

(3) The PL and a contact team.
  (a) Prepare to make physical contact with the stationary unit.
  (b) Issue a contingency plan to the PSG.
  (c) Maintain communications with the platoon and verify near and far recognition signals for linkup (good visibility and limited visibility).
  (d) Exchange far and near recognition signals with the linkup unit.
  (e) Conduct final coordination with the linkup unit.

(4) The stationary unit.
  (a) Guides the patrol from its linkup rally point to the stationary unit's linkup rally point.
  (b) Completes linkup by the time the order specifies.
  (c) The main body receives an alert before the moving unit is brought forward.

(5) The patrol continues its mission in accordance with the order.

d. Coordination checklist. The PL coordinates or obtains the following information from the unit with whom the patrol will link up.

(1) Exchange frequencies, call signs, codes, and other communications information.
(2) Near and far recognition signals.
(3) Fires coordination measures.
(4) Command relationship with the linkup unit and plan for consolidation of chain of command.
(5) Actions following linkup.
(6) Control measures such as phase lines and contact points as appropriate.

7-46. Immediately after the platoon or squad returns, personnel from higher HQ conduct a thorough debriefing. This interrogates all members of the platoon or the leaders, RTOs, and any attached personnel. Normally, the debriefing is oral but sometimes requires a written report. The following information appears on the written report:

- Size and composition of the unit conducting the patrol.
- Mission of the platoon, such as type of patrol, location, and purpose.

- Departure and return times.
- Route including checkpoints and grid coordinates for each leg or including an overlay.
- Detailed descriptions of terrain and any identified enemy positions.
- Results of any contact with the enemy.
- Unit status at the conclusion of the patrol mission including the disposition of dead or wounded Rangers.
- Conclusions or recommendations.

7-47. The objective rally point is a point out of sight, sound, and small arms' range of the objective area. Its normal location is in the direction the platoon plans to move after the completion of actions on the objective. The objective rally point is tentative until the pinpointing of the objective.

   a. <u>Occupation of the objective rally point</u> (see figure 7-6).

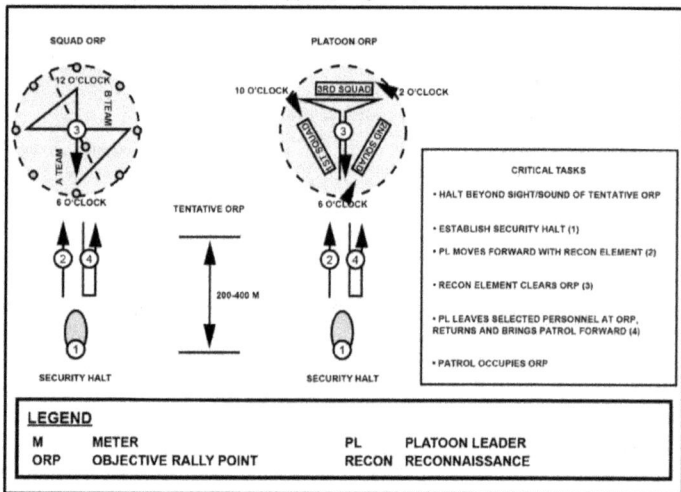

Figure 7-6. Occupation of the objective rally point

(1) The patrol halts beyond sight and sound of the tentative objective rally point (200 to 400 meters [219 to 437 yards] in good visibility; 100 to 200 meters [109 to 219 yards] in limited visibility).

(2) The patrol establishes a security halt in accordance with the unit's SOP.

(3) After issuing a five-point contingency plan to the PSG, the PL moves forward with a reconnaissance element to conduct a leader's reconnaissance of the objective rally point.

(4) For a squad-sized patrol, the PL moves forward with a compass Soldier and one member of each fire team to confirm the objective rally point

    (a) After physically clearing the objective rally point location and confirming communications with higher HQ, the PL leaves two Rangers at the 6 o'clock position, facing in opposite directions.

    (b) The PL issues a contingency plan and returns with the compass Soldier to guide the patrol forward.

    (c) The PL guides the patrol forward into the objective rally point with one team occupying from 3 o'clock through 12 o'clock to 9 o'clock and the other occupying from 9 o'clock through 6 o'clock to 3 o'clock.

(5) For a platoon-sized patrol, the PL, RTO, WSL, three ammunition bearers, a TL, a squad automatic weapon gunner, and rifle shooters join the leader's reconnaissance for the objective rally point and position themselves at 10, 2, and 6 o'clock.

    (a) The first squad in the order of march is the base squad who occupies from 10 to 2 o'clock. Their arrangement follows.

    (b) The trail squads occupy from 2 to 6 o'clock and 6 to 10 o'clock, respectively.

    (c) The patrol HQ element occupies the center of the triangle.

  b. Actions in the objective rally point.

(1) The unit prepares for the mission in the objective rally point.

(2) Once the leader's reconnaissance pinpoints the objective, the PSG generally lines up rucksacks in accordance with the unit's SOP in the center of the objective rally point.

PATROL BASE

7-48 A PB is a security perimeter set up when a squad or platoon, while conducting a patrol, halts for an extended period. Units do not occupy PBs for longer than a 24-hour period, except in an emergency. A patrol never uses the same PB twice.

  a. Typical uses.

(1) To avoid the enemy's detection by eliminating movement.

(2) To hide during a long, detailed reconnaissance.

(3) To perform maintenance on weapons and equipment and to eat and rest.

(4) To plan and issue orders.

(5) To reorganize after infiltrating an enemy area.

(6) To establish a base from which to execute several consecutive or concurrent operations.

  b. Site selection. The leader selects the tentative site from a map or by aerial reconnaissance. An effective site for a PB is easily defensible for short periods of time and away from natural lines of drift and high-speed AAs. The ideal site provides cover and concealment from both ground and air, as well as little to no tactical advantage to the enemy. Leaders confirm the site's suitability and security before the unit moves into it. Plans to establish a PB include an alternate PB site in case the first site proves unsuitable or the patrol unexpectedly evacuates.

   c. <u>Planning considerations</u>. Leaders planning a PB consider the mission and passive and active security measures. A PB's location allows the unit to accomplish its mission.

      (1) OPs and communication with OPs.

      (2) Patrol or platoon fire plan.

      (3) Alert plan.

      (4) Withdrawal plan from the PB including withdrawal routes and a rally point, rendezvous point, or alternate PB.

      (5) Security system to ensure specific Rangers are awake at all times.

      (6) Camouflage, noise, and light disciplines.

      (7) Conduct of required activities with minimum movement and noise.

      (8) Priorities of work.

   d. <u>Security measures</u>.

      (1) Select terrain the enemy probably considers of little tactical value.

      (2) Select terrain off the main lines of drift.

      (3) Select difficult terrain that impedes foot movement such as an area of dense vegetation, preferably bushes and trees spreading close to the ground.

      (4) Select terrain near a source of water.

      (5) Select terrain that is defensible for a short period and that offers good cover and concealment.

      (6) Avoid known or suspected enemy positions.

      (7) Avoid built-up areas.

      (8) Avoid ridges and hilltops except when necessary to maintain communications.

      (9) Avoid small valleys.

      (10) Avoid roads and trails.

e. Occupation (see figure 7-7).

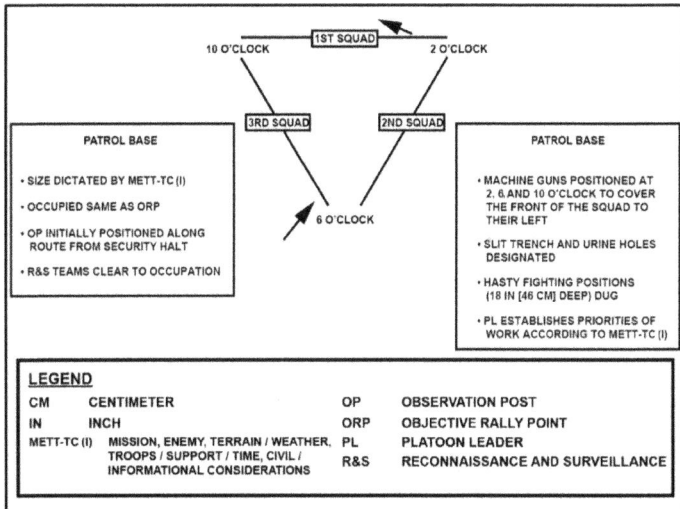

Figure 7-7. Patrol base

(1) The same manner is useful for reconnoitering and occupying a PB as for an objective rally point with the exception that platoons typically plan to enter at a 90-degree turn. The PL leaves a two-Ranger OP at the turn, and the patrol covers any tracks from the turn to the PB.

(2) The platoon moves into the PB. Squad-sized patrols generally occupy a cigar-shaped perimeter; platoon-sized patrols generally occupy a triangle-shaped perimeter.

(3) The PL and another designated leader inspect and appropriately adjust the entire perimeter.

(4) After the PL has checked each squad sector, each SL sends a two-Ranger R&S team to the PL at the command post. The PL issues the three R&S teams a contingency plan, reconnaissance method, and detailed guidance on what to look for (for example, enemy, water, built-up areas or human habitat, roads, trails, possible rally points).

(5) The PL's guidance determines where each R&S team departs. The R&S team moves a prescribed distance and direction and reenters where the PL dictates.

   (a) Squad-sized patrols do not normally send out an R&S team at night.

   (b) The R&S teams prepare a sketch of the area to the squad's front whenever possible.

   (c) The patrol remains 100-percent alert during this reconnaissance.

   (d) If the PL feels the enemy tracked or followed the patrol, the PL may elect to wait in silence at 100-percent alert before sending out the R&S teams.

   (e) The R&S teams use methods such as those referred to as the *I*, *Box*, or *T* (see figure 7-8). Regardless of the chosen method, the R&S team retains the ability to provide the PL with the same information.

   (f) Upon completion of R&S, the PL confirms or denies the PB location and either moves the patrol or begins priorities of work.

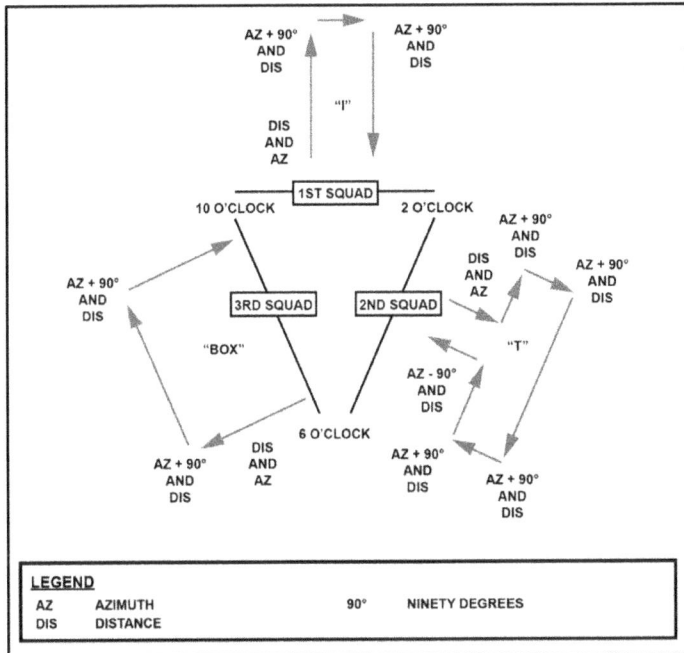

Figure 7-8. Reconnaissance and surveillance team movement methods

f. Passive (clandestine) PB (squad).
   (1) The purpose of a passive PB is for a squad-sized or smaller element to rest.
   (2) The unit moves as a whole and occupies in force.
   (3) The SL ensures the unit moves in at a 90-degree angle to the order of movement.

(4) The unit emplaces a claymore mine on the route entering the PB.

(5) The Alpha and Bravo teams sit back-to-back and facing outward to ensure at least one individual on each team is alert and providing security.

g. Priorities of work (platoon and squad). Once the R&S teams brief the PL and the PL determines the area is suitable for a PB, the leader sets or modifies defensive work priorities to establish the defense of the PB. Priorities of work are not a laundry list of tasks for completing; to be effective, priorities of work consist of a task, given time, and measurable performance standard. A clear standard for each priority of work guides the element in the successful accomplishment of each task. Whether the control of the work occurs in a centralized or decentralized manner also receives designation. A centralized manner is when the PL maintains control and all squads conduct the same priority of work. A decentralized manner is when the SLs maintain control and conduct priorities independently of one another in accordance with the METT-TC (I) variables and PL's guidance after the establishment of security. Priorities of work depend on the METT-TC (I) variables and include but are not limited to the following tasks.

(1) Security (continuous). The security plan is always the first priority of work conducted.

    (a) Prepare to use all passive and active measures to cover the entire perimeter at all times regardless of the percentage of weapons used to cover all the terrain.

    (b) Readjust after the R&S teams return or readjust based on the current priority of work such as weapons maintenance.

    (c) Employ all elements, weapons, and personnel to meet conditions of the terrain, enemy, or situation.

    (d) Assign sectors of fire to all personnel and weapons. Develop squad sector sketches and a platoon fire plan.

    (e) Confirm the locations of fighting positions for cover, concealment, observation, and fields of fire. The SLs supervise the placement of aiming stakes and claymore mines.

    (f) Use only one point of entry and exit and count personnel in and out. Challenge everyone in accordance with the unit's SOP.

    (g) Prepare hasty fighting positions at least 18 inches (45 centimeters) deep at the front and sloping gently from front to rear, with a grenade sump whenever possible.

(2) Withdrawal plan. The PL designates the signal for withdrawal, order of withdrawal, and the platoon rendezvous point or alternate PB.

(3) Communication (continuous). Maintain communications with higher HQ and OPs and within the unit. The patrol's RTOs sometimes rotate this responsibility to accomplish continuous radio monitoring maintenance, to act as runners for the PL, and to conduct other priorities of work.

(4) Mission preparation and planning. The PL uses the PB to plan, issue orders, rehearse, inspect, and prepare for future missions.

(5) Weapons and equipment maintenance. The PL ensures the maintenance of machine guns, weapon systems, communications equipment, and NVDs, as well as other equipment. Disassembly for maintenance does not occur at the same time for more than 33 percent of these items in total. Weapon disassembly never occurs at night. When one machine gun is down, raise security for all the remaining systems.

(6) Water resupply. The PSG appropriately organizes watering parties. The watering party carries canteens in an empty rucksack or duffel bag and has communications and a contingency plan prior to departure.

(7) Mess plan. At a minimum, perform security and weapons maintenance prior to mess. Normally, no more than half the platoon eats at one time. Rangers eat 1 to 3 meters (3 to 10 feet) behind their fighting positions.

(a) Rest and sleep plan management. The patrol conducts appropriate rest to prepare for future operations.

(b) Alert plan and stand-to. The PL states the alert posture and stand-to time. The plan ensures periodic checks of all positions, periodic relief of OPs, and at least one always alert leader. The patrol typically conducts stand-to at a time the unit's SOP specifies, such as 30 minutes before and after begin morning nautical twilight or end of evening nautical twilight.

(c) Resupply. Distribute or cross-load ammunition, meals, equipment, and other items.

(d) Sanitation and personal hygiene. The PSG and medic ensure the preparation and marking of a slit trench. All Rangers brush their teeth, shave, and wash their faces, hands, armpits, groins, and feet. The patrol does not leave trash.

MOVEMENT TO CONTACT

7-49. Movement to contact is one of the five types of offensive operations. Movement to contact gains or regains contact with the enemy. Once the unit makes contact, it develops the situation. Normally, a platoon conducts movement to contact as part of a larger force. The two techniques for conducting movement to contact are search and attack and cordon and search.

a. Search and attack. The search and attack technique is useful when the enemy withdraws or is dispersed, expected to avoid contact, or disengaged or when the enemy's movement in an area is denied. The search and attack technique involves multiple platoons, squads, and fire teams and coordinates their actions to make contact with the enemy. Platoons typically try to find the enemy and then fix and finish them. They combine patrolling techniques with hasty or deliberate attacks upon finding the enemy.

(1) Planning considerations.
(a) METT-TC (I) variables.
(b) Decentralized execution requirement.
(c) Mutual support requirement.
(d) Length of operations.
(e) Minimization of the Soldier's load to improve stealth and speed.
(f) Resupply and MEDEVAC.
(g) Positioning of key leaders and equipment.
(h) Employment of key weapons.
(i) PB requirement.
(j) Concept for entering the zone of action.
(k) Concept for linkups while in contact.

(2) Critical performance measures.
(a) The platoon locates the enemy without their detection.
(b) Once engaged, the platoon fixes the enemy in position and maneuvers against the enemy.
(c) The platoon maintains security throughout actions to avoid its flanking.

   b. <u>Cordon and search</u>. A technique of conducting movement to contact that isolates a target area and searches suspect locations in that target area to capture or destroy possible enemy forces and contraband. A platoon uses the cordon and search technique as part of a larger unit. A leader may task a platoon as the cordon or the search element for the company or battalion.

     (1) <u>Planning</u>. Immediately upon receipt of the mission, the company-level commander conducts reconnaissance of the target to be searched. Ideally, the battalion or brigade intelligence staff officer provides maps or satellite imagery. Leaders avoid sending a patrol to the area of the target since the patrol can unnerve the target and cause them to flee prior to the search. However, depending on how much information a commander has, no alternative may remain but to conduct reconnaissance to determine the target's location. The patrol does not necessarily alert the target when patrols frequent the area, but the patrol need not loiter in the area any longer than necessary.

     (2) <u>Preparing</u>. Rehearsals are key to successful cordon and search operations. Units develop their own requirements for what to rehearse.

     (3) <u>Organization</u> of cordon and search elements.

       (a) Outer cordon element: security and support element.

       (b) Inner cordon element: search element.

       (c) Assault force: detention and collection element.

     (4) <u>Execution</u>. As the unit approaches the objective, the inner cordon and assault forces allow enough time for the outer cordon force to set before actually arriving at the target. The locals know the sound of military vehicles, and any present subversive element tends to flee upon hearing units approach. While the impact is not always immediate, vehicles and foot traffic (aside from curious onlookers) around the objective dissipate quickly once the outer cordon is set, facilitating the movement of other elements to the objective.

TASK STANDARDS

7-50. The platoon moves NLT the time the order specifies and makes contact with the smallest element possible. The enemy does not surprise the main body. Once the platoon makes contact, it maintains contact. The platoon destroys squad-sized or smaller elements and fixes elements larger than a squad. The platoon maintains a fighting force sufficient for conducting further combat operations.

7-51. The platoon forwards reports of enemy locations and contact. When the enemy does not detect the platoon, the PL initiates a hasty attack. The platoon sustains no casualties from friendly fire. The platoon remains prepared to initiate further movement within 25 minutes of contact and account for all personnel and equipment.

Chapter 8

Battle Drills

This chapter provides select battle drills. These collective actions are designed to teach a Soldier or small unit to react and survive in common combat situations. A platoon, squad, team, or crew performs these actions when initiated to do so by a predetermined verbal or visual cue. Battle drills are performed instinctively, with little to no notice; they require minimal leader direction. Once initiated, they are vital to the success of the combat operation and critical to preserving life (see FM 7-0). Appendix A provides graphic examples. Appendix B provides quick reference cards. (See ATP 3-21.8 for information on the drills.) For the most updated battle drills, see the Army Training Network or Central Army Registry website.

React to Direct Fire Contact While Dismounted – Platoon (07-PLT-D9501)

Conditions: The platoon is conducting operations in a live training environment independently or as part of a company or larger force. The platoon is dismounted. While stationary or moving, the enemy engages the platoon with direct fire. Some iterations of this task should be performed in [mission-oriented protective posture (MOPP) 4] and at night.

Standards: The platoon reacts to direct fire contact while dismounted according to ATP 3-21.8. The squad in contact returns fire immediately and seeks cover. The squad in contact locates the enemy and places well-aimed fire on known enemy positions. Leaders point out enemy positions and identify the types of weapons, such as small arms and light machine guns. The squads not in contact assume the nearest covered and concealed position. The platoon leader (PL) reports the contact.

Cue: The drill begins when the enemy initiates direct fire contact.

TASK STEPS
(Asterisks [indicate] a leader performance step.)

1. The squad in contact immediately returns well-aimed suppressive fire on known or suspected enemy positions while taking up a covered position.
2. The squads not in contact [assume] the nearest covered and concealed position. (See figure 8-1 on page 8-2.)

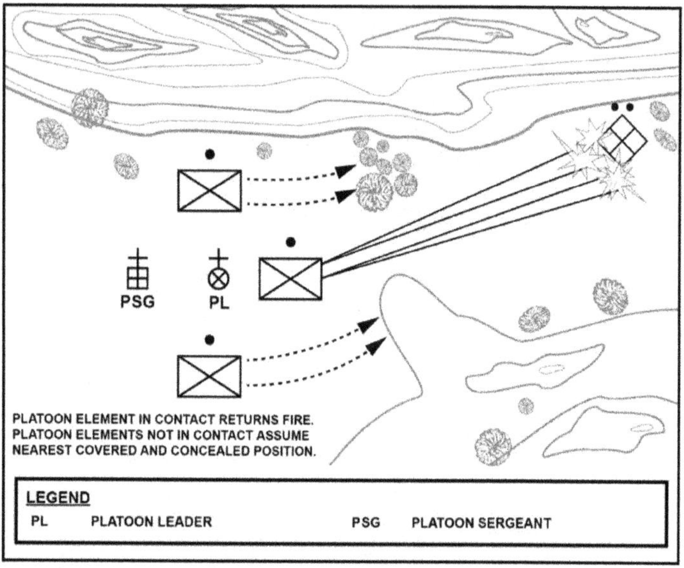

Figure 8-1. Platoon's return fire and nearest covered and concealed positions

* 3. The squad leaders engage known or suspected enemy positions with well-aimed suppressive fire and report information to the platoon leader and platoon sergeant.

* 4. The squad leaders control the fire of their Soldiers by using standard fire command (initial and supplemental) containing the following information:

    a. Alert.

    b. Weapon or ammunition (optional).

    c. Target description.

    d. Direction.

    e. Range.

f. Method.
g. Control (optional).
h. Execution.
i. Termination.
5. Soldiers maintain visual or vocal contact with their leaders and the other Soldiers on their left or right (if applicable).
6. Soldiers maintain contact with the team leader and indicate the location of enemy positions.
* 7. Leaders visually or vocally check the status of their personnel.
* 8. The squad leaders maintain visual contact with the platoon leader.
* 9. The platoon leader moves up to a covered and concealed position where best to observe, communicate, and control the engagement. (See figure 8-2.)

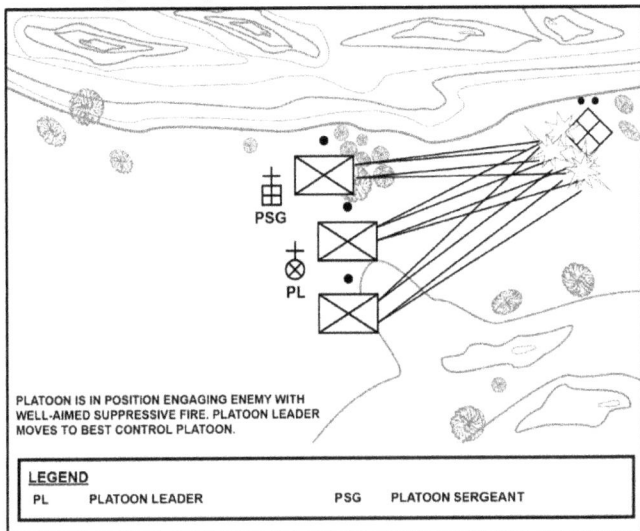

Figure 8-2. Well-aimed fires and platoon leader control

    a. The platoon leader brings the radio-telephone operator, forward observer, squad leader of the nearest squad, and one crew-served weapon team (machine gun team if available).

    b. The squad leaders of the squads not in contact move to the front of their squad.

    c. The platoon sergeant moves forward with the remaining crew-served weapons and links up with the platoon leader and assumes control of the support element.

* 10. The platoon leader determines whether or not the platoon can gain and maintain suppressive fires with the squad already in contact (based on the volume and accuracy of enemy fires against the squad in contact).

* 11. The platoon leader confirms the [commander's] criteria to disengage and determines whether or not the platoon must move out of the engagement area.

* 12. The platoon leader makes an assessment of the situation and identifies [the]—

    a. Location of the enemy position and obstacles.

    b. Size of the enemy force engaging the squad in contact. (The number of enemy automatic weapons, the presence of any vehicles, and the employment of indirect fire are indicators of enemy strength.)

    c. Vulnerable flanks.

    d. Covered and concealed flanking routes to the enemy positions.

* 13. The platoon leader decides whether to conduct an assault, bypass (if authorized by the company commander), or break contact.

* 14. The platoon leader reports the situation and begins to maneuver the platoon.

React to Direct Fire Contact While Dismounted – Squad (07-SQD-D9501)

Conditions: The squad is conducting operations in a live training environment independently or as part of a platoon or larger force. The squad is dismounted. While stationary or moving, the enemy engages the squad with direct fire. Some iterations of this task should be performed in MOPP 4 and at night.

Standards: The squad reacts to direct fire contact while dismounted according to ATP 3-21.8. The team in contact returns fire immediately and seeks cover. The team in contact locates the enemy and places well-aimed fire on known enemy positions. Leaders point out enemy positions and identify the types of weapons, such as small arms and light machine guns. The team not in contact [assumes] the nearest covered and concealed position. The squad leader (SL) reports the contact.

Cue: The drill begins when the enemy initiates direct fire contact.

TASK STEPS
(Asterisks [indicate] a leader performance step.)

    1. The team in contact immediately returns well-aimed suppressive fire on known or suspected enemy positions while taking up a covered position.

    2. The team not in contact assumes the nearest covered and concealed position. (See figure 8-3.)

Figure 8-3. Squad's return fire and nearest covered and concealed position

TEAM IN CONTACT RETURNS FIRE.
TEAM NOT IN CONTACT ASSUMES
NEAREST COVERED AND
CONCEALED POSITION.

* 3. The team leaders engage known or suspected enemy positions with well-aimed suppressive fire and report information to the squad leader.
* 4. The team leaders control the fire of their teams by using standard fire command (initial and supplemental) containing the following information:
  a. Alert.
  b. Weapon or ammunition (optional).
  c. Target description.
  d. Direction.
  e. Range

  f. Method.
  g. Control (optional).
  h. Execution.
  i. Termination.
5. Soldiers maintain visual or vocal contact with their team leader and the other Soldiers on their left or right (if applicable).
6. Soldiers maintain contact with the team leader and indicate the location of enemy positions.
* 7. Leaders visually or vocally check the status of their personnel.
* 8. The team leaders maintain visual contact with the squad leader.
* 9. The squad leader moves up to a covered and concealed position where best to observe, communicate, and control the engagement. (See figure 8-4.)

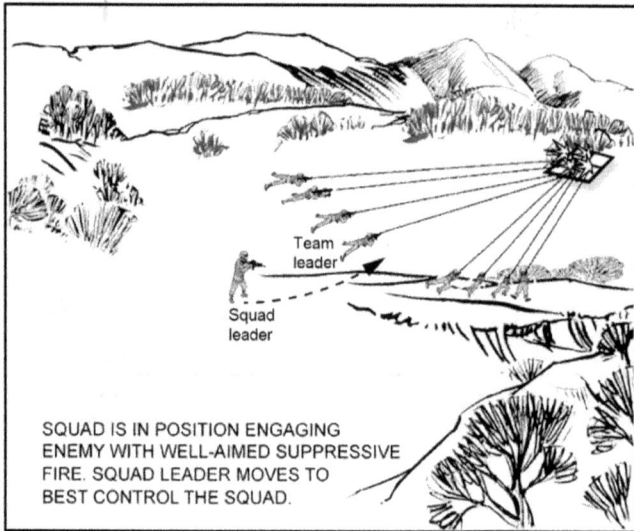

SQUAD IS IN POSITION ENGAGING ENEMY WITH WELL-AIMED SUPPRESSIVE FIRE. SQUAD LEADER MOVES TO BEST CONTROL THE SQUAD.

Figure 8-4. Well-aimed fires and squad leader control

* 10. The squad leader determines whether or not the squad can gain and maintain suppressive fires with the team already in contact (based on the volume and accuracy of enemy fires against the team in contact).
* 11. The squad leader confirms the [commander's] criteria to disengage and determines whether or not the squad must move out of the engagement area.
* 12. The squad leader makes an assessment of the situation and identifies [the]—
  a. Location of the enemy position and obstacles.
  b. Size of the enemy force engaging the team in contact. (The number of enemy automatic weapons, the presence of any vehicles, and the employment of indirect fires are indicators of enemy strength.)
  c. Vulnerable flanks.
  d. Covered and concealed flanking routes to enemy positions.
* 13. The squad leader decides whether to conduct an assault, bypass (if authorized by the platoon leader), or break contact.
* 14. The squad leader reports the situation and begins to maneuver the squad.

Conduct a Platoon Assault (07-PLT-D9514)

Conditions: The platoon is conducting operations in a live training environment independently or as part of a company or larger force. The platoon is part of a movement to contact or an attack. The enemy initiates direct fire contact on the lead squad. Some iterations of this task should be performed in MOPP 4 and at night.

Standards: The platoon conducts an assault according to ATP 3-21.8. The platoon locates and suppresses the enemy, establishes supporting fire, and assaults the enemy position using fire and maneuver. The platoon destroys or causes the enemy to withdraw and conducts consolidation and reorganization.

Cue: The drill begins when the enemy initiates direct fire contact.

TASK STEPS
(Asterisks [indicate] a leader performance step.)

1. The platoon conducts action on enemy contact, as follows.
   a. The squad or section in contact reacts to contact by immediately returning well-aimed fire on known enemy positions.
   b. The squad or section leader, in contact, notifies the platoon leader of the action.
   c. The squad or section, in contact, attempts to achieve suppressive fires.
   d. Dismounted Soldiers assume the nearest covered positions.
   e. Vehicles ([Bradley Fighting Vehicles], Strykers, and [high mobility multipurpose wheeled vehicles]), if applicable, move out of the beaten zone and the engagement area.
* 2. The platoon leader gives the command to dismount the vehicles, if mounted.
3. The platoon sergeant takes control of the vehicles, if applicable.
* 4. The platoon leader, [radio-telephone] operator, [forward observer (FO)], squad leader of the next squad, and one machine gun team move forward to [link up] with the squad leader of the squad or section in contact.
* 5. The squad leader of the trail squad moves to the front of the trail squad's lead fire team.

* 6. The weapons squad leader and second machine gun team move forward and [link up] with the platoon leader. If directed, the weapons squad leader assumes control of [the] base-of-fire element and positions the machine guns to add suppressive fires against the enemy.

7. The platoon sergeant repositions vehicles, as necessary, to provide observation and supporting fire against the enemy.

* 8. The platoon leader assesses the situation.

9. The squad or section in contact cannot achieve suppressive fire. The squad or section in contact takes the following actions:

   a. Reports limited capabilities to effectively suppress enemy forces to the platoon leader.

   b. Establishes a base of fire:

      (1) The squad leader deploys the squad to provide effective, sustained fires on the enemy position.

      (2) The squad leader reports the final position to the platoon leader.

10. The remaining squads, not in contact, take up covered and concealed positions in place[and] observe to the flanks and rear of the platoon.

* 11. The platoon leader determines if the squad in contact can gain suppressive fire against the enemy based on the volume and accuracy of the enemy's return fire[.]

   a. The platoon leader directs the squad, with one or both machine guns, and vehicle element in contact to continue suppressing the enemy:

      (1) The squad in contact destroys or suppresses enemy weapons that are firing most effectively against it.

      (2) The vehicle section in contact destroys or suppresses enemy weapons that were firing most effectively against them, including vehicles and crew-served weapons.

      (3) The squad in contact places screening smoke (M203 and M320) to prevent the enemy from seeing the maneuver element.

   b. The platoon leader determines additional actions required and deploys another squad, [a] second vehicle section, and the second machine gun team to suppress the enemy position.

Note. The platoon leader may direct the trail leader to position this squad and vehicle section[and the] weapons squad leader to position one or both machine gun teams in a better support-by-fire position.

---

* 12. The platoon leader, again, determines if the platoon can gain suppressive fires against the enemy:

   a. The platoon leader continues to suppress the enemy with the two squads, two machine guns, and vehicle-mounted weapons.

      (1) The trail squad leader assumes control of the base-of-fire element (squad in contact, machine gun teams, and any other squads designated by the platoon leader).

      (2) The platoon sergeant assumes control of the vehicle section and base-of-fire element (squad in contact and machine gun teams designated by the platoon leader).

      (3) The platoon FO calls for and adjusts indirect fire based on the platoon leader's directions.

Note. The platoon leader does not wait for indirect fires before continuing [with actions].

---

   b. The platoon leader determines additional [actions required] and deploys the last squad to provide flank and rear security and reports the situation to the company commander.

   c. The platoon continues to suppress or fix the enemy with direct and indirect fire[and] responds to orders from the company commander.

13. The platoon assaults the enemy position. If the squads in contact together with the machine guns and vehicle element can suppress the enemy, the platoon leader determines if the remaining squads that are not in contact can maneuver:

   a. The platoon leader makes the following assessment:

     (1) Location of enemy positions and obstacles.

     (2) Size of [the] enemy force. (The number of enemy automatic weapons, the presence of any vehicles, and the employment of indirect fires are indicators of enemy strength.)

     (3) Vulnerable flank.

     (4) Covered and concealed flanking route to the enemy position.

   b. The platoon leader maneuvers the squads into the assault:

     (1) Once the platoon leader has ensured that the base-of-fire element is in position and providing suppressive fires, [the platoon leader leads] the assaulting squads to the assault position.

     (2) If the vehicle section can effectively suppress the enemy element, the platoon leader may reposition the weapons squad or machine gun to an intermediate or local support-by-fire position to provide additional suppression during the assault.

     (3) Once in position, the platoon leader gives the prearranged signal for the base-of-fire element to lift or shift direct fires to the opposite flank of the enemy position. (The assault element MUST pick up and maintain effective fires throughout the assault. Handover of responsibility for direct fires from the base-of-fire element to the assault element is critical.)

     (4) The platoon FO shifts indirect fires to isolate the enemy position.

     (5) The assaulting squad(s) fight through enemy positions using fire and maneuver. The platoon leader controls the movement of the squads, assigns specific objectives for each squad, and designates the main effort or base maneuver element. The base-of-fire element must be able to identify the near flank of the assaulting squad(s).

     (6) In the assault, the squad leader determines the way in which to move the elements of the squad based on the volume and accuracy of enemy fire against the squad and the amount of cover afforded by the terrain. In all cases, each Soldier uses individual movement techniques, as appropriate.

       (a) The squad leader designates one fire team to support the movement of the other team by fires.

       (b) The squad leader designates a distance or direction for the team to move and accompanies one of the fire teams.

       (c) Soldiers must maintain contact with team members and leaders.

       (d) Soldiers time their firing and reloading in order to sustain their rate of fire.

       (e) The moving fire team proceeds to the next covered position. Teams use the wedge formation when assaulting. Soldiers move in rushes or by crawling.

       (f) The squad leader directs the next team to move.

      (g) If necessary, the team leader directs Soldiers to bound forward as individuals within buddy teams. Soldiers coordinate their movement and fires with each other within the buddy team[and] maintain contact with their team leader.

      (h) Soldiers fire from covered positions. They select the next covered position before moving and rush forward (no more than five seconds), or [they] use high or low crawl techniques based on terrain and enemy fires.

  c. [When the] platoon leader determines the assaulting squad(s) cannot continue to move, the platoon leader deploys the squad(s) to suppress the enemy and reports to the company commander. The platoon continues suppressing enemy positions and responds to the orders of the company commander.

14. The platoon consolidates on the objective once the assaulting squads have seized the enemy positions by:
  a. Establishing local security.
  b. The platoon leader signals for the base-of-fire element to move up into designated positions.
  c. The platoon leader assigns sectors of fire for each squad and vehicle.
  d. The platoon leader positions key weapons and vehicles to cover the most dangerous avenue(s) of approach.
  e. The platoon sergeant begins coordination for ammunition resupply.
  f. Soldiers take hasty defensive positions.
  g. The platoon leader and FO develop a quick fire plan.
  h. Placing out observation posts to warn of enemy counterattacks.

15. The platoon reorganizes by:
  a. Reestablishing the chain of command.
  b. Redistributing and resupplying ammunition.
  c. [Staffing] crew-served weapons first.
  d. Redistributing critical equipment such as radios, [night-vision devices; and protection from chemical, biological, radiological, and nuclear exposure].
  e. Treating casualties and evacuating wounded.
  f. Filling vacancies in key positions.
  g. Searching, silencing, segregating, safeguarding, and speeding [enemy prisoners of war] to collection points.
  h. Collecting and reporting enemy information and materiel.

* 16. The platoon leader sends a [situation report] to the company commander.

Conduct a Squad Assault (07-SQD-D9515)

Conditions: The squad is conducting operations in a live training environment independently or as part of a platoon or larger force. The squad is part of a movement to contact or an attack. The enemy initiates direct fire contact. Some iterations of this task should be performed in MOPP 4 and at night.

Standards: The squad conducts an assault according to ATP 3-21.8. The squad locates and suppresses the enemy, establishes supporting fire, and assaults the enemy positions using fire and maneuver. The squad destroys [the enemy] or causes the enemy to withdraw and conducts consolidation and reorganization.

Cue: The drill begins when the enemy initiates direct fire contact.

TASK STEPS
(Asterisks [indicate] a leader performance step.)

1. The fire team in contact immediately returns well-aimed fire on the known enemy position and calls out the direction and distance of the enemy.

Note. Soldiers receiving fire take up the nearest positions that afford cover and concealment.

---

* 2. The squad leader reports contact to the platoon leader.
3. The fire team in contact takes the following actions:
   a. The team leader directs fires using tracers or standard fire commands.
   b. Fire team Soldiers in contact move to positions (bound or crawl) where they can fire their weapons, position themselves to ensure that they have observation, field of fire, cover, and concealment.
   c. Fire team Soldiers continue to fire and report known or suspected enemy positions to the fire team leader.
4. The fire team not in contact takes covered and concealed positions in place and observes the flanks and rear of the squad.
* 5. The squad leader moves to a position to observe the enemy and assess the situation and then [takes] the following actions:
   a. Reports the enemy size and location, and any other information to the platoon leader.
   b. Requests through higher, immediate indirect fire support.
   c. Determines if the fire team in contact can gain suppressive fire based on the volume and accuracy of the enemy fire.
6. The fire team in contact suppresses the enemy and the following actions take place:
   a. The team leader identifies target reference points with the squad leader for shifting fires.
   b. The fire team becomes the support by fire team.
   c. The fire team continues suppression of the enemy as follows.
      (1) The fire team leader continues to control fires using tracers or standard fire commands. Fires must be well-aimed and continue at a sustained rate with no lulls.
      (2) The fire team destroys or suppresses enemy crew-served weapons first.
      (3) The fire team places smoke (M203/320) on the enemy position to obscure it.
      (4) Buddy teams alternate their fire so that both are not reloading their weapons at the same time.
* 7. The squad leader moves to the fire team not in contact[.]

Note. The fire team not in contact becomes the assault team.

---

   a. [The squad leader links] up with the fire team leader not in contact.
   b. [The squad leader identifies] the fire team not in contact as the assault team.
* 8. The squad leader maneuvers the assault fire team into the assault position and the following actions take place:
   a. The squad leader adjusts fires (both direct and indirect) based on the rate of the assault element movement and the minimum safe distances of weapon systems.

    b. The squad leader directs the forward observer, if available, to shift indirect fire, including smoke, to isolate the enemy position.

    c. The squad leader gives the prearranged signal for the support by fire team to shift direct fire to the opposite flank of the enemy position.

    d. The assault fire team assumes and maintains effective fire throughout the assault.

    e. The support by fire team confirms the shift of direct fires with the prearranged signal.

9. The assault team fights through the enemy position using fire and maneuver and takes the following actions:

    a. The assault team leader controls the movement of the team.

    b. The assault team conducts fire and maneuver based on [the] volume and accuracy of enemy fires and the amount of cover afforded by the terrain as follows:

        (1) Assault team leader designates a distance and direction for the assault team and moves with that element.

        (2) Assault team leader directs Soldiers to move as individuals or teams.

        (3) Soldiers maintain contact with team members and leaders.

        (4) Soldiers move using 3- to 5-second rushes or the low or high crawl techniques, taking advantage of available cover and concealment.

        (5) Soldiers time their firing and reloading to sustain their rate of fire.

10. The support by fire team maintains visual contact of the near flank of the assaulting team as the assaulting team conducts the assault.

11. If the assault element cannot continue to move, the squad leader positions the squad to suppress the enemy and reports to higher headquarters.

12. The squad consolidates and reorganizes as follows:

    a. The squad establishes local security.

    b. The squad leader signals for the support by fire team to move into designated positions.

    c. The squad places out observation posts to warn of enemy counterattacks.

    d. The squad leader assigns sectors of fire for each element.

    e. The Soldiers establish hasty fighting positions.

    f. The squad leader positions key weapons to cover the most dangerous avenue of approach.

    g. The squad leader develops a quick fire plan.

    h. The squad reestablishes the chain of command.

    i. The squad [staffs] crew-served weapons and then individual weapons.

    j. The squad leader begins coordination of ammunition resupply.

    k. The squad leader consolidates ammunition, casualties and equipment reports.

    l. The squad redistributes and resupplies ammunition.

    m. The squad redistributes critical equipment such as radios, [night-vision devices, and protection from chemical, biological, radiological, and nuclear exposure].

    n. The squad treats and [evacuates] wounded.

    o. The squad searches, silences, segregates, safeguards, speeds, and tags detainees.

* 13. The squad leader reports situation to the platoon leader.

Break Contact – Platoon (07-PLT-D9505)

Conditions: The platoon is conducting operations in a live training environment independently or as part of a company or larger force. The platoon is moving as part of a larger force, conducting a movement to contact or an attack. Following direct fire contact with the enemy, the platoon leader decides to break contact. Indirect fires are available. Some iterations of this task should be performed in MOPP 4 and at night.

Standards: The platoon breaks contact according to ATP 3-21.8. Using fire, movement and maneuver, the platoon continues to move until the enemy cannot observe or place fire on them. The platoon leader reports the contact.

Cue: The drill begins when the platoon leader orders the platoon to break contact.

TASK STEPS
(Asterisks [indicate] a leader performance step.)

* * 1. While receiving direct fire from the enemy or on orders, the platoon leader orders the platoon to break contact.
* * 2. The platoon leader directs one squad or fire team to suppress by fire to support the disengagement of the remainder of the platoon.
* * 3. The platoon leader orders a distance and direction, terrain feature, or last rally point of the movement of the platoon element in contact.

Note. This distance should not exceed small arms range to ensure supporting fire.

---

* * 4. The platoon leader employs direct fires to suppress enemy positions. (See figure 8-5 on page 8-14.)

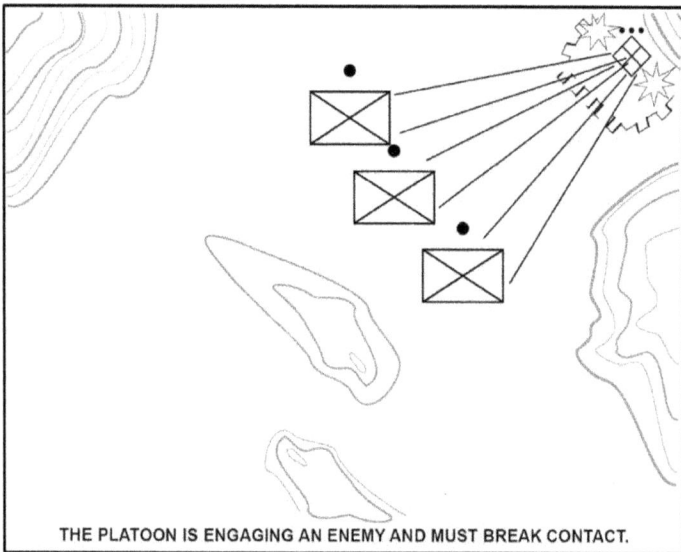

THE PLATOON IS ENGAGING AN ENEMY AND MUST BREAK CONTACT.

Figure 8-5. Platoon's direct and indirect fires to suppress enemy

5. The moving element moves to occupy the overwatch position, employs smoke (M203/M320, smoke grenades, indirect fires, and other options) to screen movement.

> *Note.* If necessary, [the bounding element] employs fragmentation and concussion grenades to facilitate breaking contact.

6. The base-of-fire element continues to suppress the enemy.

7. The moving element occupies their overwatch position and engages enemy positions. (See figure 8-6.)

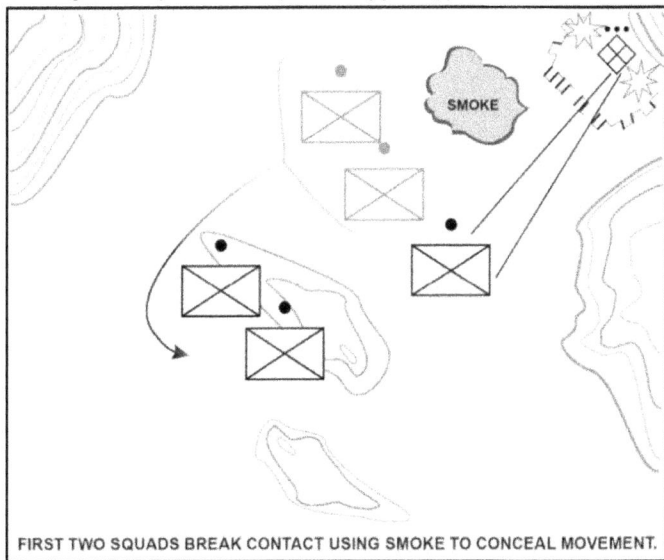

FIRST TWO SQUADS BREAK CONTACT USING SMOKE TO CONCEAL MOVEMENT.

Figure 8-6. Moving element's overwatch and enemy engagement

" 8. The platoon leader directs the base-of-fire element to move to its next covered and concealed position. (See figure 8-7.)

Note. Based on the terrain[and the] volume and accuracy of the enemy's fire, the moving element may need to use fire and movement techniques.

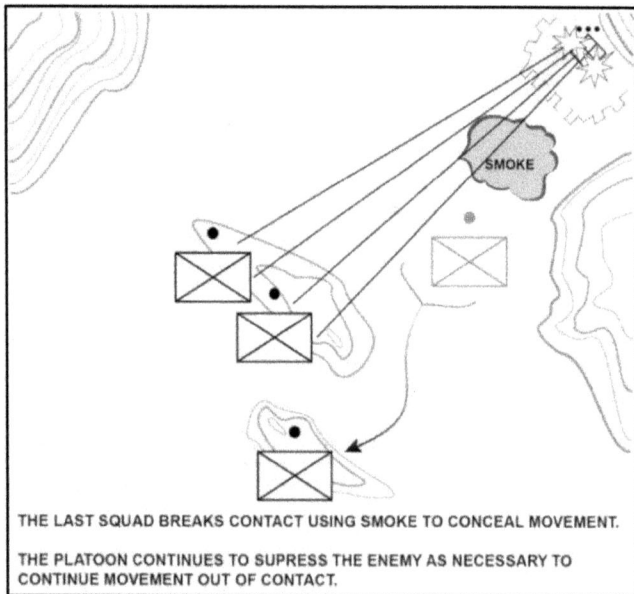

THE LAST SQUAD BREAKS CONTACT USING SMOKE TO CONCEAL MOVEMENT.

THE PLATOON CONTINUES TO SUPRESS THE ENEMY AS NECESSARY TO CONTINUE MOVEMENT OUT OF CONTACT.

Figure 8-7. Fire and movement techniques

[9.] The platoon continues to move away from the enemy until:
  a. It breaks contact (the platoon must continue to suppress the enemy as it breaks contact).
  b. Its elements are in the assigned positions to continue the mission.
* 10. Leaders account for Soldiers, report the situation, reorganize as necessary, and continue the mission.
* 11. The platoon leader moves the platoon onto an azimuth or alternate route away from enemy forces.

Note. The platoon leader should consider changing the unit's direction of movement once contact is broken. This reduces the ability of the enemy to place effective indirect fire on the platoon.

12. Squads and Teams that become divided stay together and move to the last designated rally point.

Break Contact – Squad (07-SQD-D9505)

Conditions: The squad is conducting operations in a live training environment independently or as part of a platoon or larger force. The squad is moving as part of a larger force, conducting a movement to contact or an attack. Following direct fire contact with the enemy, the squad leader decides to break contact. Some iterations of this task should be performed in MOPP 4 and at night.

Standards: The squad breaks contact according to ATP 3-21.8. Using fire and movement, the squad continues to move until the enemy cannot observe or place fire on them. The squad leader reports the contact.

Cue: The drill begins when the squad leader orders the squad to break contact.

TASK STEPS
(Asterisks [indicate] a leader performance step.)

* 1. While receiving direct fire from the enemy or on orders, the squad leader orders the squad to break contact.
* 2. The squad leader directs one team to suppress by fire to support the disengagement of the remainder of the squad.
* 3. The squad leader orders a distance and direction, terrain feature, or last rally point of the movement of the team in contact.

Note. This distance should not exceed small arms range to ensure supporting fire.

" 4. The squad leader employs direct fires to suppress enemy positions. (See figure 8-8.)

SQUAD IS ENGAGING ENEMY
AND MUST BREAK CONTACT.

Figure 8-8. Squad's direct fire to suppress enemy

5. The moving team moves to occupy the overwatch position, employs smoke (M203/M320, smoke grenades, and other options) to screen movement.

Note. If necessary, [the moving team] employs fragmentation and concussion grenades to facilitate breaking contact.

6. The base-of-fire team continues to suppress the enemy.
7. The moving team occupies their overwatch position and engages enemy positions. (See figure 8-9.)

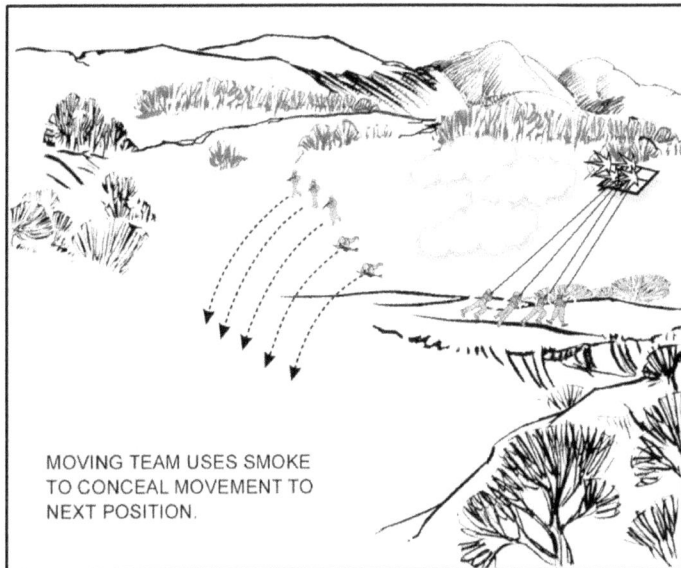

MOVING TEAM USES SMOKE TO CONCEAL MOVEMENT TO NEXT POSITION.

Figure 8-9. Moving team's overwatch and enemy engagement

* 8. The squad leader directs the base-of-fire team to move to its next covered and concealed position. (See figure 8-10.)

Note. Based on the terrain[and the] volume and accuracy of the enemy's fire, the moving team may need to use fire and movement techniques.

TEAM MOVES INTO NEXT
COVERED AND CONCEALED
POSITION AND SUPPRESSES ENEMY.
UNIT CONTINUES TO SUPPRESS AND
BOUND.

Figure 8-10. Moving team's fire and movement techniques

9. The squad continues to move away from the enemy until:
    a. It breaks contact (the squad must continue to suppress the enemy as it breaks contact).
    b. Its teams are in the assigned positions to continue the mission.
* 10. Leaders account for Soldiers, report the situation, reorganize as necessary, and continue the mission.
* 11. The squad leader moves the squad onto an azimuth or alternate route away from enemy forces.

Note. The squad leader should consider changing the unit's direction of movement once contact is broken. This reduces the ability of the enemy to place effective indirect fire on the squad.

---

12. Teams and Soldiers that become divided stay together and move to the last designated rally point.

React to Ambush (Dismounted) – Platoon (07-PLT-D9502)

Conditions: The platoon is conducting operations in a live training environment independently or as part of a company or larger force. The platoon is moving tactically dismounted in close terrain. The platoon moves into an [enemy-prepared] kill zone. The enemy initiates contact with the most casualty-producing weapon or detonation of explosives and a high volume of well-aimed fire from covered and concealed positions. Indirect fire is available. Some iterations of this task should be performed in MOPP 4 and at night.

Standards:
The platoon reacts to an ambush according to ATP 3-21.8.

If a Near Ambush, then Soldiers in the kill zone immediately return fire on known or suspected enemy positions and assault through the kill zone. Soldiers not in the kill zone locate and place well-aimed suppressive fire on the enemy. The platoon assaults through the kill zone and destroys the enemy.

If a Far Ambush, then Soldiers in the kill zone immediately return fire on known or suspected enemy positions and suppress the enemy. Soldiers not in the kill zone assault the enemy using fire and movement. The platoon assaults through the kill zone and destroys the enemy.

Cue: The drill begins when the enemy initiates an ambush.

TASK STEPS
(Asterisks [indicate] a leader performance step.)

1. The platoon is moving dismounted, receives a high volume of well-aimed fire from the enemy, and takes the following actions (see figure 8-11 on page 8-22).

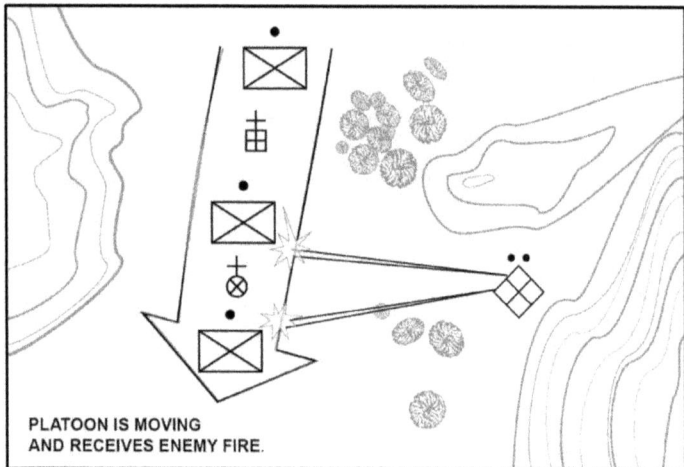

Figure 8-11. Enemy fire upon platoon

a. React to a Near Ambush in which the enemy is within hand grenade range:

  (1) The teams/squads in the kill zone execute one of the following two actions:

     (a) Return fire immediately, and if cover is not available, without order or signal immediately assume the prone position, and throw smoke grenades.

     (b) Return fire immediately and, if cover is available, without order or signal, occupy the nearest covered position, immediately assume the prone position, and throw smoke grenades. (See figure 8-12.)

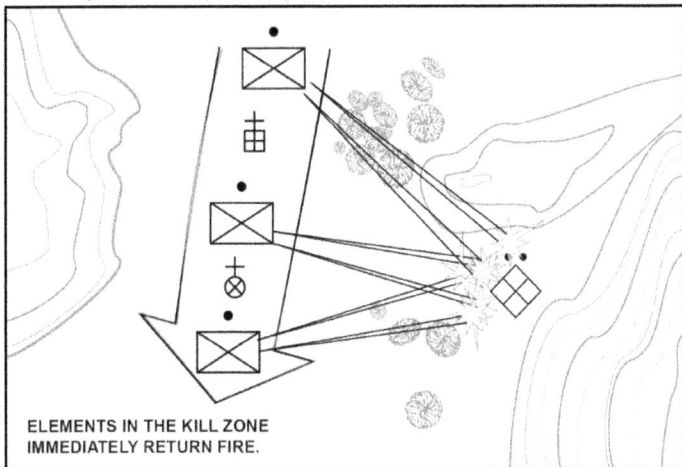

ELEMENTS IN THE KILL ZONE IMMEDIATELY RETURN FIRE.

Figure 8-12. Platoon's immediate return fire

(2) The teams/squads in the kill zone, immediately after the explosion of the smoke grenades, assault through the ambush position using fire and movement.

(3) The teams/squads not in the kill zone identify the enemy location, place well-aimed suppressive fire on the enemy's position, and shift fire as Soldiers assault the objective.

(4) The Soldiers/team in the kill zone continue to assault through and destroy the enemy position. (See figure 8-13.)

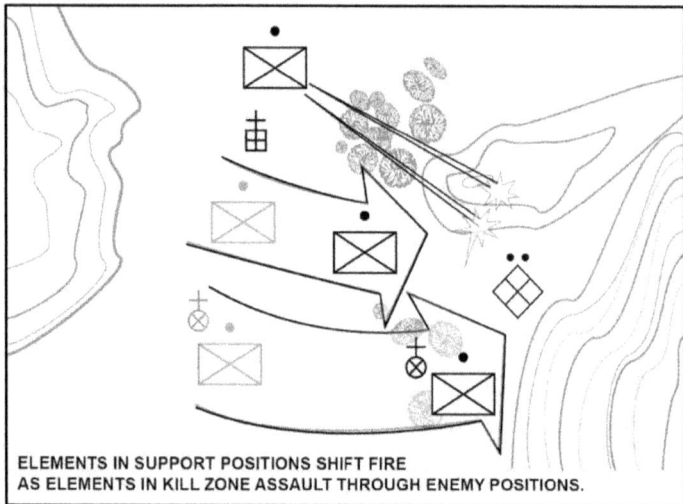

ELEMENTS IN SUPPORT POSITIONS SHIFT FIRE
AS ELEMENTS IN KILL ZONE ASSAULT THROUGH ENEMY POSITIONS.

Figure 8-13. Platoon's assault through enemy positions

b. React to a Far Ambush in which the enemy is beyond hand grenade range:

(1) Teams/squads receiving fire immediately return fire, seek cover, and place well-aimed suppressive fire on the enemy's position.

(2) [The squad leader or platoon leader leads the teams/squads] not receiving fire along a covered and concealed route to the enemy's flank to assault the enemy using fire and movement.

(3) Teams/squads in the kill zone shift suppressive fires as the assaulting Soldiers fight through and destroy the enemy.

(4) The platoon leader calls for and adjusts indirect fire according to [the METT-TC (I) variables: mission, enemy, terrain and weather, troops and support available, time available, civil considerations, and informational considerations].

* 2. The platoon leader reports the contact.

React to Ambush (Dismounted) – Squad (07-SQD-D9502)

Conditions: The squad is conducting operations in a live training environment independently or as part of a platoon or larger force. The squad is moving tactically dismounted in close terrain. The squad moves into an [enemy-prepared] kill zone. The enemy initiates contact with the most casualty-producing weapon or detonation of explosives and a high volume of well-aimed fire from covered and concealed positions. Some iterations of this task should be performed with MOPP 4 and at night.

Standards:
The squad reacts to an ambush according to ATP 3-21.8.

If a Near Ambush, then Soldiers in the kill zone immediately return fire on known or suspected enemy positions and assault through the kill zone. Soldiers not in the kill zone locate and place well-aimed suppressive fire on the enemy. The squad assaults through the kill zone and destroys the enemy.

If a Far Ambush, then Soldiers in the kill zone immediately return fire on known or suspected enemy positions and suppress the enemy. Soldiers not in the kill zone assault the enemy using fire and movement. The squad assaults through the kill zone and destroys the enemy.

Cue: The drill begins when the enemy initiates an ambush.

TASK STEPS
(Asterisks [indicate] a leader performance step.)

1. The squad is moving dismounted, receives a high volume of well-aimed enemy fire, and takes the following actions (see figure 8-14 on page 8-26.)

Figure 8-14. Enemy fire upon squad

a. React to a Near Ambush in which the enemy is within hand grenade range:

(1) The Soldiers/team in the kill zone execute one of the following two actions:

(a) Return fire immediately, and if cover is not available, without order or signal immediately assume the prone position, and throw smoke grenades.

(b) Return fire immediately and, if cover is available, without order or signal, occupy the nearest covered position, immediately assume the prone position, and throw smoke grenades. (See figure 8-15.)

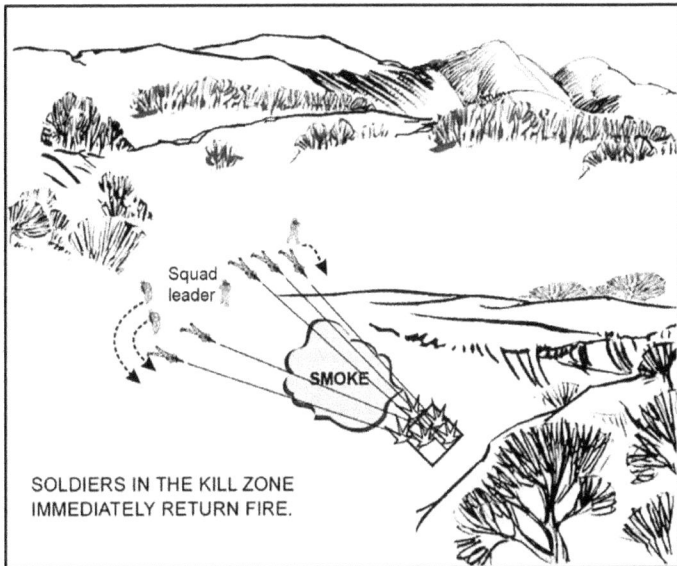

Figure 8-15. Squad's immediate return fire

(2) The Soldiers/team in the kill zone, immediately after the explosion of the smoke grenades, assault through the ambush position using fire and movement.

(3) The Soldiers/team not in the kill zone identify the enemy location, place well-aimed suppressive fire on the enemy's position, and shift fire as Soldiers assault the objective.

(4) The Soldiers/team in the kill zone continue to assault through and destroy the enemy position. (See figure 8-16 on page 8-28.)

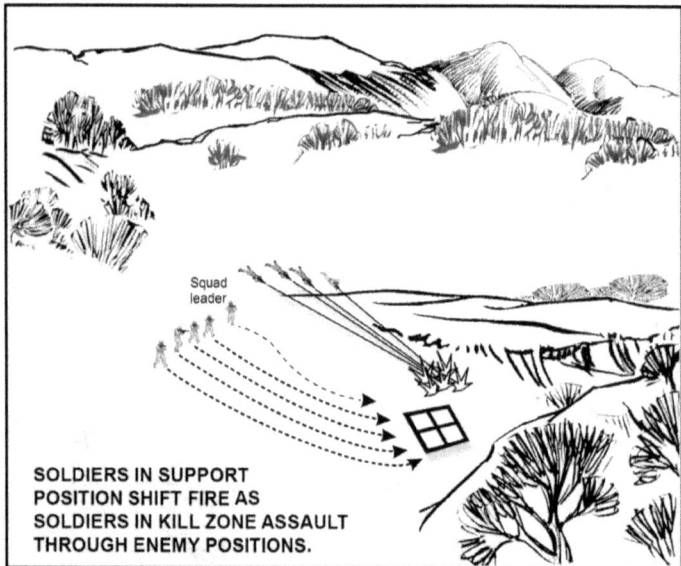

SOLDIERS IN SUPPORT
POSITION SHIFT FIRE AS
SOLDIERS IN KILL ZONE ASSAULT
THROUGH ENEMY POSITIONS.

Figure 8-16. Squad's assault through enemy positions

    b. React to a Far Ambush in which the enemy is beyond hand grenade range.

        (1) Soldiers/team receiving fire immediately return fire, seek cover, and place well-aimed suppressive fire on the enemy's position.

        (2) [The team leader or squad leader leads the Soldiers/team] not receiving fire along a covered and concealed route to the enemy's flank to assault the enemy using fire and movement.

        (3) Soldiers/team in the kill zone shift suppressive fire as the assaulting Soldiers fight through and destroy the enemy.

* 2. The squad leader reports the contact.

> # DANGER
>
> Do **not** attempt to cook off or milk any grenade type. Cooking off or milking the safety lever before employment can result in personnel death or serious injury.
>
> When walls or floors are thin or damaged and grenades are in play, falling or exploding fragments can kill or seriously injure Soldiers.

Enter and Clear a Room – Squad (07-SQD-D9509)

Conditions: The squad is conducting operations in a live training environment as part of a platoon or larger force. The squad receives an order to clear a room. Enemy personnel are believed to be in the building. Noncombatants may be present in the building and are possibly intermixed with the enemy personnel. The squad has support and security elements positioned at the initial foothold and outside the building. Some iterations of this task should be performed MOPP 4 and at night.

Standards: The squad enters and clears a room according to ATP 3-21.8. The squad clears and secures the room by killing or capturing the enemy while minimizing friendly casualties, noncombatant casualties, and collateral damage. The squad complies with the rules of engagement (ROE), maintains a sufficient fighting force to repel an enemy counterattack, and continues operations.

Cue: This drill begins on the order of the squad leader to enter and clear a room.

TASK STEPS
(Asterisks [indicate] a leader performance step.)

* 1. The squad leader—
   a. Occupies a position to best control the security and clearing teams.
   b. Directs a security team to secure corridors or hallways outside the room with appropriate firepower.
   c. Identifies the entry point.
   d. Gives the clearing team leader an order to clear the room.

> *Note.* If the squad is conducting high-intensity operations and grenades are being used, the unit must comply with the ROE and consider the building structure. A Soldier of the clearing team prepares at least one grenade (fragmentation, concussion, or stun grenade), throws the grenade into the room, and announces: FRAG OUT or STUN OUT. The use of grenades should be consistent with the ROE and building structure. Soldiers can be injured from fragments if walls and floors are thin or damaged.

* 2. The team leader (normally the number two Soldier) does the following:
    a. Takes a position to best control the clearing team outside the room.
    b. Determines the method of entry into the room.
    c. Gives the signal to enter and clear the room.
3. The first two Soldiers enter the room with the second Soldier immediately after the first Soldier as follows: (See figure 8-17.)

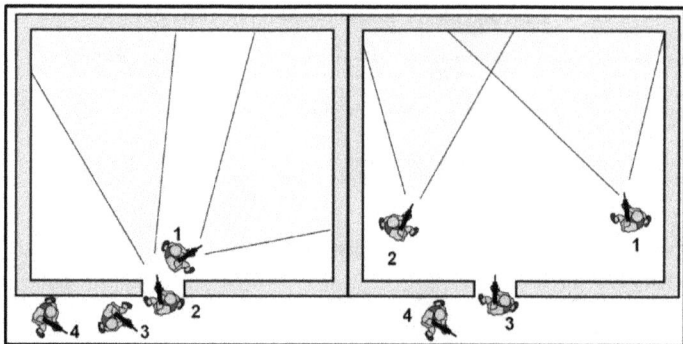

Figure 8-17. Clearing a room—first Soldier immediately followed by second Soldier

    a. The first Soldier enters the room, moves left or right along the path of least resistance to one of two corners, and assumes a position of domination facing into the room. During movement, the Soldier scans the sector and eliminates all immediate threats.
    b. The second Soldier (normally the clearing team leader) enters the room immediately after the first Soldier and moves in the opposite direction of the first Soldier to a point of domination. During movement, the Soldier eliminates all immediate threats in the sector.

Note. During high-intensity combat, the Soldiers enter immediately after the grenade detonates. Both Soldiers enter while firing aimed bursts into their sectors, engaging all threats or hostile targets to cover their entry.

If the first or second Soldier discovers the room is small or a short room (such as a closet or bathroom), [the Soldier] announces SHORT ROOM or SHORT. The clearing team leader informs the third and fourth Soldiers whether or not to stay outside the room or to enter.

4. The third Soldier moves in the opposite direction of the second Soldier while scanning and clearing the sector and assuming the point of domination. (See figure 8-18.)

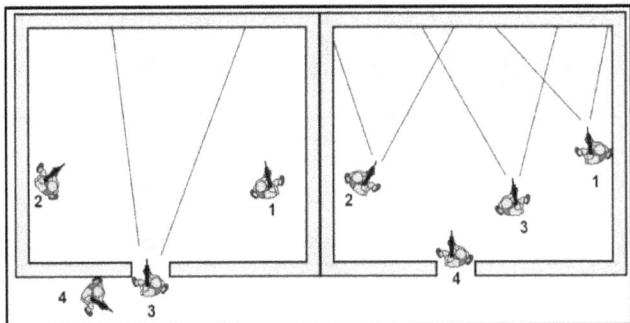

Figure 8-18. Clearing a room—third Soldier enters room and clears sector

5. The fourth Soldier moves opposite of the third Soldier to a position dominating a sector. (See figure 8-19.)

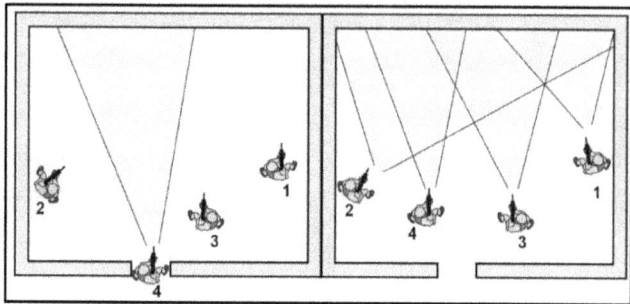

Figure 8-19. Clearing a room–fourth Soldier enters room and clears sector

6. All Soldiers engage enemy combatants with precision-aimed fire and identify noncombatants to avoid collateral damage.

Note. If necessary or on order, number one and number two Soldiers of the clearing team may move deeper into the room while overwatched by the other team members.

---

* 7. The clearing team leader scans and clears the room by—
    a. Ensuring all threats are neutralized.
    b. Ensuring all noncombatants are secured.
    c. Announcing, CLEAR to the squad leader when the room is clear.
    d. Establishing security.
* 8. The squad leader directs the security team to continue to secure the corridor or hallway.
* 9. The squad leader enters the room and—
    a. Makes a quick assessment of the room and threat.
    b. Determines if the squad has firepower to continue clearing their assigned sector.
    c. Reports to the platoon leader that the room is clear.
    d. Requests needed sustainment to continue clearing the squad's sector.
    e. Marks entry point according to the unit [standard operating procedure].
10. The squad consolidates and reorganizes, as needed.

React to Indirect Fire While Dismounted – Platoon (07-PLT-D9504)

Conditions: The platoon is conducting operations in a live training environment independently or as part of a company or larger force. The platoon is dismounted. While stationary or moving, a member of the platoon alerts, INCOMING or a round impacts nearby. Some iterations of this task should be performed in MOPP 4 and at night.

Standards: The platoon reacts to indirect fire while dismounted according to ATP 3-21.8. Soldiers immediately seek the best available cover. The platoon moves out of the area to the designated rally point after the impacts. The platoon leader reports the contact.

Cue: The drill begins when any platoon member alerts, INCOMING or a round impacts nearby

TASK STEPS
(Asterisks [indicate] a leader performance step.)

1. A member of the platoon hears or observes artillery impacting near the platoon and alerts the platoon with the announcement, INCOMING!
2. Soldiers immediately assume the prone position or immediately move to nearby available cover during initial impacts. (See figure 8-20.)

Figure 8-20. Soldiers' reaction to indirect fire

* 3. The platoon leader orders the platoon to move to a rally point by giving a direction and distance.
4. Soldiers move rapidly in the direction and distance to the designated rally point and reestablish security after the impact.

5. Leaders regain accountability of their Soldiers, weapons, and equipment at the rally point and treat casualties as required.

* 6. The platoon leader reports the contact.

React to Indirect Fire While Dismounted – Squad (07-SQD-D9504)

Conditions: The squad is conducting operations in a live training environment independently or as part of a platoon or larger force. The squad is dismounted. While stationary or moving, a member of the squad alerts, INCOMING, or a round impacts nearby. Some iterations of this task should be performed in MOPP 4 and at night.

Standards: The squad reacts to indirect fire while dismounted according to ATP 3-21.8. Soldiers immediately seek the best available cover. The squad moves out of the area to the designated rally point after the impacts. The squad leader reports the contact.

Cue: The drill begins when any squad member alerts, INCOMING, or a round impacts nearby.

TASK STEPS
(Asterisks [indicate] a leader performance step.)

1. A member of the squad hears or observes artillery impacting near the squad and alerts the squad with the announcement, INCOMING!
2. Soldiers immediately assume the prone position or immediately move to nearby available cover during initial impacts. [See figure 8-20 on page 8-33.]
* 3. The squad leader orders the squad to move to a rally point by giving a direction and distance.
4. Soldiers move rapidly in the direction and distance to the designated rally point and reestablish security after the impacts.
5. Leaders regain accountability of their Soldiers, weapons, and equipment at the rally point and treat casualties as required.
* 6. The squad leader reports the action.

Chapter 9

Military Mountaineering

In the mountains, commanders face the challenge of maintaining their units' combat effectiveness and efficiency. (See TC 3-97.61 for information on military mountaineering.)

TRAINING AND PLANNING

9-1. Military mountaineering training provides units with tactical mobility in otherwise inaccessible mountainous terrain. Rangers receive training in the fundamental mobility and climbing skills units need to move safely and efficiently in mountainous terrain.

9-2. Rangers who are to conduct combat operations in a mountainous environment receive extensive training to prepare them for the rigor of mountain operations. Some of the areas follow:

- Mountainous environment characteristics.
- Basic mountaineering equipment care and use.
- Mountain bivouac techniques.
- Mountain communications.
- Mountain travel and walking techniques.
- Mountain navigation, hazard recognition, and route selection.
- Rope management and knots.
- Natural and artificial anchors.
- Belay and rappel techniques.
- Installation construction and use such as rope bridges.
- Rock climbing fundamentals.
- CASEVAC raising and lowering systems.
- Individual movement on snow and ice.
- Mountain stream crossings including water survival techniques.

9-3. Unique planning considerations for mountain operations include movement and insertion techniques, methods of reducing the enemy's ability to observe, identification of likely enemy positions, and possible ambush and contact locations in the development of movement techniques, routes, and actions on contact in compartmentalized and canalized terrain.

9-4. Planning also takes into consideration possible degraded reaction times from other ground units and challenges for sustainment functions (for example, resupply, MEDEVAC, CASEVAC). Plans also factor in possible climate changes to weather, precipitation, wind, and temperature, which can occur quickly and have extreme impact.

DISMOUNTED MOBILITY

9-5. One requirement of Ranger military mountaineers is to assess a vertical obstacle and develop a COA to overcome the obstacle. They also possess the skills to accomplish the plan. Assessment of a vertical obstacle requires understanding of route classifications and the levels of difficulty they represent. Without a solid understanding of the difficulty of a chosen route, mountain leaders can place their lives and the lives of other Soldiers in extreme danger. Ignorance is the most dangerous hazard in the mountainous environment.

9-6. Mountain operations require physically fit Soldiers and leaders experienced in operations in this terrain. Problems arise in moving personnel and transporting loads up and down steep and varied terrain to accomplish the mission. Chances for success in this environment increase when leaders have experience operating in the same conditions as their troops. Acclimatization, conditioning, and training are important to successful military mountaineering. Table 9-1 is useful when deciding the platoon's movement formation and techniques during the planning process.

Table 9-1. Terrain classification considerations

CLASS	TERRAIN	MOBILITY	PLANNING CONSIDERATIONS	REQUISITE SKILL LEVEL
1	Gentle slopes or trails	Walking	No special training; general environmental acclimatization only.	None
2	Steeper, rugged	Walking with some required use of hands	Recommended environmental acclimatization. Unit movement, standard operating procedure, or battle drill training on steep terrain.	Basic mountaineering helpful but not required
3	Easy to climb or scramble over	Easy climbing with fixed ropes where exposed or risking a fall	Environmental acclimatization. Soldier load management. Unit movement, standard operating procedure, or battle drill training on steep terrain. Unit movement on fixed lines.	Basic mountaineering to place simple fixed ropes and installations
4	Steep, exposed	Requiring fixed ropes	Extensive environmental acclimatization. Soldier load management. Unit movement, standard operating procedure, battle drill training on steep terrain. Unit movement on fixed lines. Negotiation of near vertical obstacles. Route selection.	Basic mountaineering plus assault climbing to establish anchors, fixed ropes, and hauling systems

Table 9-1. Terrain classification considerations (continued)

CLASS	TERRAIN	MOBILITY	PLANNING CONSIDERATIONS	REQUISITE SKILL LEVEL
5	Near vertical to vertical	Technical climbing	Extensive environmental acclimatization. Extensive Soldier load management. Assault climbing. Technical rope rescue. Rope ascending and descending.	Recommended assault climbing to advise commanders and supervise complex rope systems

TASK ORGANIZATION

9-7. Leaders take special consideration when task-organizing a Ranger platoon in a mountainous environment. Leaders plan for such specific teams as CASEVAC teams, reconnaissance and installation teams, and security teams specific to any potentially encountered obstacle. Leaders also consider who are the strong swimmers and climbers and what climbing-level-qualified mountaineers are within the platoon.

9-8. Upon completion of task organization, the platoon organizes and consolidates all rescue and mountaineering equipment.

RESCUE EQUIPMENT

9-9. Rangers use a litter system that functions like a standard basket-type litter but is more compact, lightweight, and versatile. The stretcher is made of low-density polyethylene plastic with solid brass grommets, nylon webbing, and steel buckles.

9-10. The system not including taglines weighs 19 pounds (8.6 kilograms) when packed inside the carrying case. When packed, the case is 9 inches (23 centimeters) in diameter and 36 inches (91 centimeters) in length. The system withstands extreme temperatures including extremely cold weather (down to minus 120 degrees Fahrenheit [49 degrees Celsius]) and does not begin to melt until 450 degrees Fahrenheit (232 degrees Celsius).

9-11. When possible, link up with the aviation support unit to discuss the unit's SOPs and conduct rehearsals. Components of the litter system include the following:
- Stretcher.
- Nylon backpack.
- Horizontal lift slings with 10,000-pound (4,536-kilogram) tensile strength.
- Vertical lift sling with 5,800-pound (2,631-kilogram) tensile strength.
- Locking, steel carabiner with 9,000-pound (4,082-kilogram) tensile strength.
- Tow strap with 300-pound (136-kilogram) tensile strength.
- Four webbing handles with 300-pound (136-kilogram) tensile strength.

9-12. The litter system has different loading procedures.

    a. Unpack and unroll the rescue litter.
    b. Bend the litter backward and in half to make it lie flat.
    c. Place a patient in the litter by one of two methods.
       (1) For the logroll method, load a casualty when the stabilization of the patient's entire body is critical.
          (a) Place the litter next to the patient.
          (b) Roll the patient onto one side and slide the litter as far under the patient as possible.
          (c) Roll the patient onto the litter and carefully slide the patient into the litter's center.
          (d) Secure the patient to the litter.
       (2) For the slide method, load a casualty when the injuries prevent rolling the patient onto one side and when confined spaces prevent the use of the logroll method.
          (a) Place the foot end of the stretcher at the patient's head.
          (b) One person straddles the stretcher and supports the patient's head, neck, and shoulders.
          (c) Two people grab straps and pull the stretcher under the patient while slightly lifting the patient's head and shoulders.

9-13. Carefulness protects the casualty from potential environmental injuries. Practices include wrapping the patient in a space blanket and sleeping bag during cold weather operations and monitoring potential overheating in hot weather operations. The procedure to fasten straps and buckles follows.
    a. Lift the sides of the stretcher and fasten the straps to their complementary buckles.
    b. Feed the foot straps through the unused buckles at the foot of the stretcher and fasten to the buckles.
    c. After placing and strapping the patient securely in the litter, lace the stretcher with the 30-foot lift sling of kernmantle rope. The lacing provides additional security to the litter, and the fixed loop created by the double figure eight knot serves as the point of attachment for lifting or lowering the stretcher.

9-14. A vertical lift is useful on sloping terrain and when the canopy prevents a horizontal lift. When rigging this lift, remember this type of lift forces the weight of the casualty's body onto the lower extremities. The procedure to make a vertical lift follows.
    a. Create a fixed loop in the middle of the rope by tying a double figure eight knot.
    b. Pass the tails through the grommet on either side of the patient's head and snug the knot against the stretcher.
    c. Feed the ropes through the grommets along the sides, pass through the handles and through the grommets at the foot end of the stretcher, and secure with a square knot.
    d. Route the pigtails through the lower carrying handles (outside to inside) and secure the ends. The litter is towable by the tow handle strap hauling system. It is ready for lifting or lowering.

9-15. When rigging for a horizontal lift, remember the head strap is 4 inches (10 centimeters) shorter than the foot strap. The horizontal lift position is preferable for lifting or lowering. The correct rigging and closing of the litter render it impossible for the casualty to fall accidentally from the stretcher.

9-16. Horizontal is the most stable platform for the casualty since the stretcher absorbs all the casualty's weight. Control taglines are also easily employable and much more efficient, reducing spin in horizontal lifts. Lifting horizontally requires the use of two (head and foot) lift straps rated at 9,000 pounds each, thereby creating redundant attaching points. The head strap is 4 to 6 inches (10 to 15 centimeters) shorter than the foot strap to ensure the patient's head remains slightly raised. The procedure to make a horizontal lift follows.
    a. Insert one end of the head strap through the slot at the head end of the stretcher.
    b. Route the strap under the stretcher and then through the slot on the opposite side.

    c. Repeat with the foot strap at the foot end of the stretcher

    d. Equalize the weight on all the straps and insert one steel carabiner through the sewn loops on all four straps.

    e. Ensure the removal of the horizontal lift straps whenever the stretcher is to be dragged. This prevents damage to the straps.

9-17. Upon the completion of proper rigging, the stretcher is ready to lift. The procedure to ascend vertical terrain with a casualty follows.

    a. Secure the casualty in a stretcher for carrying and dragging

    b. Secure the casualty in a stretcher for horizontal and vertical helicopter evacuation.

    c. Task-organize a platoon for moving a casualty to include a carrying squad, security squads, machine guns, and key leaders. The PL focuses on the entire tactical situation while controlling the platoon. Rotate the carrying squad when they are moving the casualty over a long distance. The PSG controls the CASEVAC.

    d. Establish the primary anchor with a sling rope and two opposite and opposed carabiners and the secondary anchor for the six- to eight-wrap Prusik safety.

    e. Have teams move ahead to set up anchors that expedite moving the casualty up multiple pitches.

9-18. The procedure to descend vertical terrain with a casualty follows.

    a. Lower the casualty on a Munter hitch (see paragraph 9-40) with a six- to eight-wrap Prusik safety (see paragraph 9-42).

    b. Everyone else rappels down, and the last person configures a retrievable rappel.

    c. Teams move down and establish anchors to expedite the lowering when there are multiple pitches or rope lengths.

MOUNTAINEERING EQUIPMENT

9-19. Mountaineering equipment refers to all the parts and pieces that allow the trained Ranger to accomplish many tasks in the mountains. The importance of this gear to the mountaineer is no less than that of the rifle to the Infantry Soldier.

9-20. Army mountaineering kits comprise three separate but integrated kits with state-of-the-art, commercial equipment that meets the highest industry standards. The separate kits enable the commander to tailor the equipment to the mission's environment.

    a. The high angle mountaineering kit is designed for a minimally trained Infantry brigade combat team platoon (40 personnel) moving on rope installations that assault climbers emplaced through steep terrain void of ice or snow. This kit provides each Soldier in the platoon with a harness, locking and nonlocking carabiners, sewn webbing runners, 7-millimeter accessory cord, and a belay/rappel device. There are also static installation ropes, a rope cutter, and a rope washer.

    b. A trained assault climber team (three personnel) uses the assault climber team kit to establish rope installations that minimally trained Soldiers using the high angle mountaineering kit then traverse. The assault climber team kit provides each Soldier of the assault climber team with a harness, locking and nonlocking carabiners, sewn webbing runners, mechanical ascenders, chock pick, assault climber bag, 7-millimeter accessory cord, and a belay/rappel device. There are also dynamic climbing ropes and rock protection equipment including spring-loaded camming devices and chocks.

    c. An Infantry platoon trained in techniques for operating in steep terrain covered by snow or ice uses the snow and ice mobility kit. This kit provides each Soldier in the platoon with an avalanche transceiver, crampons, ice axe, and snowshoes. Included are also avalanche shovels, probes, and ice and snow anchors.

ROPES AND CORDS

9-21. Ropes and cords are the most important pieces of mountaineering equipment, and proper selection deserves careful thought. These items are a lifeline in the mountains, so selecting the right type and size is of utmost importance. All ropes and cord used in mountaineering and climbing today possess the same basic configuration. This is the kernmantle construction technique, which is essentially a woven sheath protecting a core of nylon fibers—similar to parachute cord.

9-22. Ropes come in two types: static and dynamic. Their types refer to their ability to stretch under tension. A static rope has very little stretch, perhaps as little as 1 to 2 percent, and is most useful in rope installations. A dynamic rope is most useful for climbing and general mountaineering. Its ability to stretch up to one-third of its overall length makes it the right choice any time the user is at risk of falling. Dynamic and static ropes come in various diameters and lengths. For most military applications, a standard 10.5- or 11-millimeter by 50-meter (164-foot) dynamic rope and 11-millimeter by 45-meter (148-foot) static rope is sufficient.

9-23. A short section of static rope or static cord is called a sling rope or cordelette. These are critical pieces of personal equipment in mountaineering operations. The diameter usually ranges from 7 to 8 millimeters and the length extends up to 21 feet (6.5 meters). 8 millimeters by 15 feet (4.5 meters) is the minimum Ranger standard.

9-24. Cordage or small-diameter rope is indispensable to the mountaineer. Its many uses make it a valuable piece of equipment. All cord is static, and its construction is of the same manner as that of larger rope.

9-25. Rope used daily remains employed no longer than 1 year. An occasionally used rope generally remains employed up to 5 years with proper care (as follows).

    a. Inspect ropes thoroughly before, during, and after use for cuts, frays, abrasions, mildew, and soft or worn spots.
    b. Never step on a rope or unnecessarily drag it on the ground.
    c. Avoid running a rope over sharp or rough edges; pad as necessary.
    d. Keep ropes away from oils, acids, and other corrosive substances.
    e. Avoid running ropes across one another under tension. For example, nylon-to-nylon contact damages ropes.
    f. Do not leave ropes knotted or under tension longer than necessary.
    g. Clean in cool water, loosely coil, and hang to dry out of direct sunlight. Ultraviolet light harms synthetic fibers. When the rope is wet, hang it to drip-dry at room temperature on a rounded wooden peg. Do not apply heat.

Webbing and Slings

9-26. Loops of tubular webbing or cord, called slings or runners, are the simplest pieces of equipment and some of the most useful. Their uses are seemingly endless, the most critical being their use as the link between the climber, rope, carabiners, and anchors. Manufacturers make runners predominantly from 9/16-inch or 1-inch tubular webbing and tie or sew them.

9-27. The carabiner is one of the most versatile pieces of equipment available in the mountains. This simple piece of gear is the critical connection between the climber, rope, and protection that is attaching the climber to the mountain. Carabiners must be strong enough to hold hard falls yet light enough for the climber to carry a quantity of them easily. Today's highly technical metal alloys allow carabiners to meet both these requirements. Steel carabiners are still widely in use by the military, but those made of lighter and stronger materials are replacing them. Basic carabiner construction affords several different shapes.

9-28. *Protection* is the generic term for a piece of equipment (natural or artificial) that is useful for constructing an anchor. Protection unites with a climber, belayer, and climbing rope to form the lifeline of the climbing team. The rope connects two climbers, and the protection connects them to the rock or ice. Figure 9-1 depicts removable artificial protection such as stoppers and spring-loaded camming devices. Figure 9-2 on page 9-8 depicts fixed (usually permanent) artificial protection.

Figure 9-1. Examples of traditional (removable) protection used on rocks

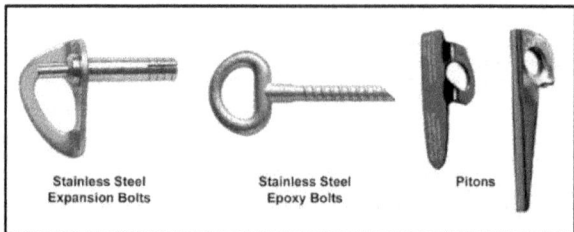

Figure 9-2. Examples of fixed (permanent or semipermanent) protection used on rocks

Equipment Inspection

9-29. Ropes require inspection before, during, and after each use, especially when employed around loose rock or sharp edges. Although the core of kernmantle rope is not visible, it can sustain damage separate from and without damage to the sheath. Check kernmantle rope by carefully inspecting the sheath before and after use while coiling the rope. When coiling, be aware of how the rope feels as it runs through the hands. Immediately note and tie off any palpable lumps or depressions. Carabiners and hardware also require inspection before, during, and after each use. The inspection looks for opening gates to open and close freely and for the absence of rust and corrosion. A fine file is useful for smoothing sharp edges or burrs.

Anchors

9-30. Anchors are the base for all installations and roped mountaineering techniques. Anchors must be strong enough to support the entire weight of the loads or impacts they take. The incorporation of several pieces of artificial or natural protection together composes one multipoint anchor. Anchors are classified *artificial* or *natural.*

    a. <u>Artificial anchor</u> construction uses all artificial materials. The most common anchors incorporate traditional or fixed protection. (See figure 9-3.)

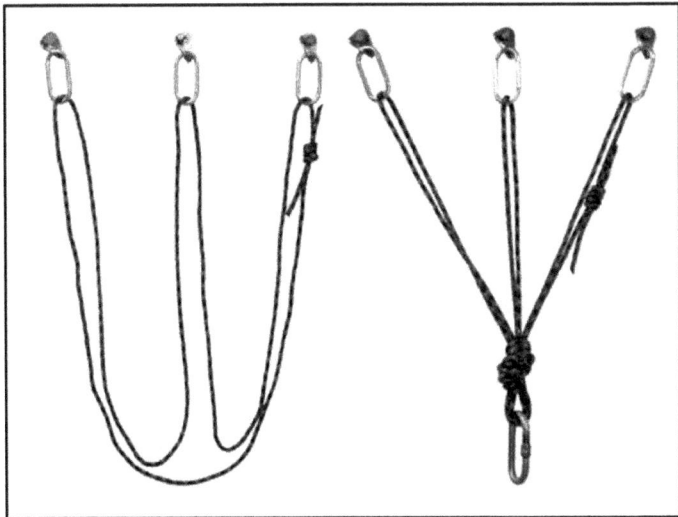

Figure 9-3. Example of a three-point, pre-equalized anchor using fixed artificial protection

   b. <u>Natural anchors</u> are usually very strong and often simple to construct using minimal equipment. Trees, shrubs, and boulders are the most common. All natural anchors require naught more than a simple method of attaching a rope. Regardless of the type of natural anchor used, the anchor must be strong enough to support the entire weight of the load.

      (1) <u>Trees</u> are probably the most widely used of all anchors. In rocky terrain, trees usually have a very shallow root system. Check this by pushing or tugging on the tree to see how well rooted it is. Anchor as low as possible to prevent excess leverage on the tree. Use padding on soft, sap-producing trees to keep sap off ropes and slings.

   c. <u>Rock projections and boulders</u> must be heavy enough and have a stable enough base to support the load.

  d. If no more suitable anchor is available than <u>bushes or shrubs</u>, route a rope around the bases of several bushes. As with trees, place the anchoring rope as low as possible to reduce leverage on the anchor. Ensure all the vegetation is healthy and well rooted to the ground.

  e. A <u>tensionless anchor</u> is useful on high-load installations such as bridging. The wraps of the rope around the anchor (see figure 9-4) absorb the tension of the installation and keep the tension off the knot and carabiner. Tie it with a minimum of four wraps around the anchor, and note that a smooth anchor such as a small tree, pipe, or rail sometimes requires several more wraps. Wrap the rope from top to bottom. Place a fixed loop into the end of the rope and loosely attach it back onto the rope with a carabiner.

Figure 9-4. Example of tensionless, natural anchor

KNOTS

9-31. Proficiency with knots and rope is vitally important for Rangers, especially in mountaineering situations. Familiarity with the terminology associated with knots and rope in general is critical. (See figure 9-5.) These terms include the following:

- *Running or working end* — the loose, or working, end of the rope.
- *Standing end* — the static, stationary, or nonworking end of the rope.
- *Bight* — an open eyelet of rope created by placing the running end alongside the standing end.
- *Loop* — a closed eyelet of rope created by placing the running end across the standing end.
- *Overhand knot* — formed by creating an eyelet and inserting the free running end through it.

- *Half hitch* — an overhand knot tied around an object with the pigtail pulled perpendicular to the standing end
- *Pigtail* — the portion of the running end of the rope between the safety knot and the end of the rope.
- *Turn* — formed by passing the running end of a rope 360 degrees around an object.
- *Round turn* — formed by wrapping the rope around an object 1-1/2 times. A round turn is useful for distributing the load over a small-diameter anchor (3 inches [8 centimeters] or narrower). It is also useful around larger diameter anchors to reduce the tension on the knot or provide additional friction.
- *Dress* — the proper arrangement of all the knot parts, removing unnecessary kinks, twists, and slack so all the rope parts of the knot make contact.

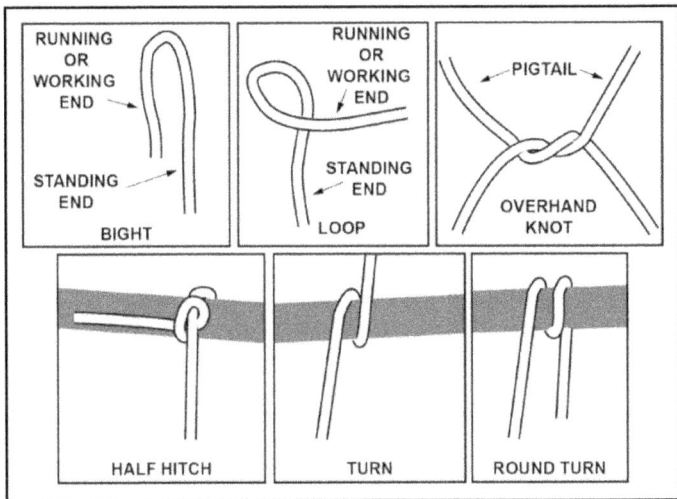

Figure 9-5. Rope terms

9-32. A <u>square knot</u> (see figure 9-6) joins two ropes of equal diameter. The knot has two interlocking bights; the running ends exit on the same side of the standing portion of rope. An overhand knot on the standing end secures each tail. When dressing the knot, leave at least a 4-inch (10-centimeter) tail on the working end.

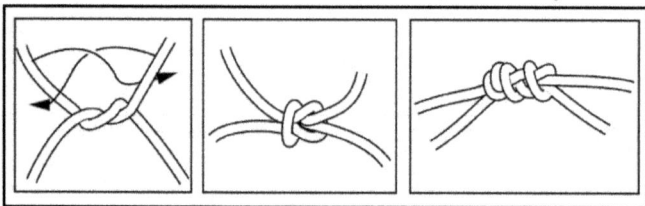

Figure 9-6. Square knot

9-33. A <u>round turn with two half hitches</u> (see figure 9-7) is a constant tension anchor knot. The rope forms a complete turn around the anchor point (earning the knot its name, round turn) with both ropes parallel and touching but not crossing. Both half hitches are tightly dressed against the round turn with the locking bar on top. When dressing the knot, leave at least a 4-inch (10-centimeter) tail on the working end.

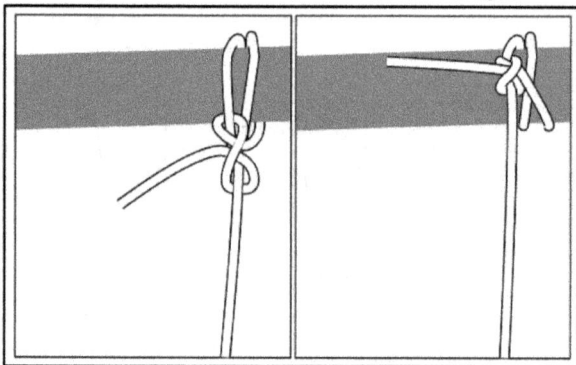

Figure 9-7. Round turn with two half hitches

9-34. To make a double figure eight knot (see figure 9-8), use a figure eight loop knot to form a fixed loop in the end of the rope. Tie it at the end of the rope or anywhere along the length of the rope. Two ropes parallel to each other in the shape of a figure eight, with no twists in the figure eight, form a figure eight loop knot. Fixed loops are large enough to insert a carabiner. When dressing the knot, leave at least a 4-inch (10-centimeter) tail on the working end.

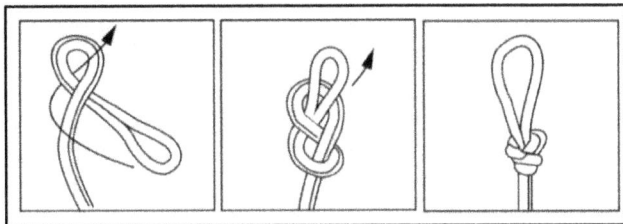

Figure 9-8. Double figure eight knot

9-35. The end-of-the-rope clove hitch (see figure 9-9) is an intermediate anchor knot that requires constant tension. Make two turns around the anchor. A locking bar runs diagonally from one side to the other. Leave no more than one rope's width between turns. The locking bar is opposite the direction of pull. When dressing the knot, leave at least a 4-inch (10-centimeter) tail on the working end.

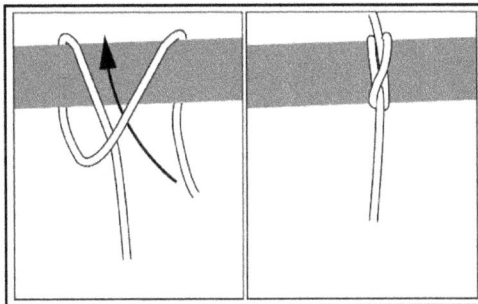

Figure 9-9. End-of-the-rope clove hitch

9-36. The <u>middle-of-the-rope clove hitch</u> (see figure 9-10) secures the middle of a rope to an anchor. The knot forms two turns around the anchor. A locking bar runs diagonally from one side to the other. Leave no more than one rope's width between turns. The locking bar is opposite the direction of pull.

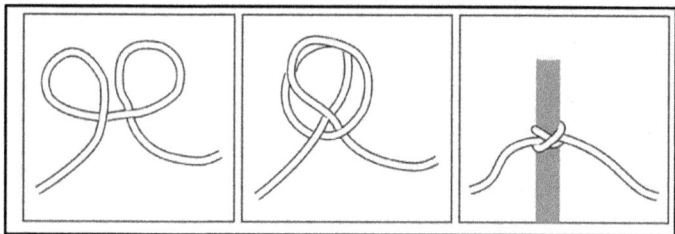

Figure 9-10. Middle-of-the-rope clove hitch

9-37. The <u>rappel seat</u> is a rope harness for rappelling and climbing. Tie it with the left or right hand. The leg straps do not cross and are tightly centered on the buttocks (see figure 9-11, step 3). The leg straps form locking half hitches on the rope around the waist (see figure 9-11, step 3). Tie the square knot on the right hip (see figure 9-11, step 4) and finish with two overhand knots (see figure 9-11, step 5). Insert the carabiner around all the ropes with the opening gate opening up and away (see figure 9-11, step 6). The carabiner does not come in contact with the square knot or overhand knot. The rappel seat is tight enough to prevent the inserting of a fist between the person and the harness.

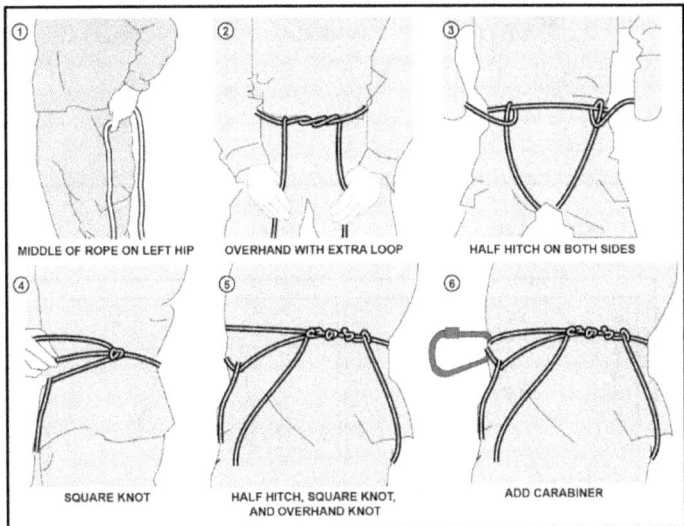

Figure 9-11. Rappel seat

9-38. The rerouted figure eight knot (see figure 9-12) is an anchor knot that also attaches a climber to a climbing rope. Form a figure eight in the rope and pass the working end around an anchor. Reroute the end back through to form a double figure eight. Tie the knot with no twists. When dressing the knot, leave at least a 4-inch (10-centimeter) tail on the working end.

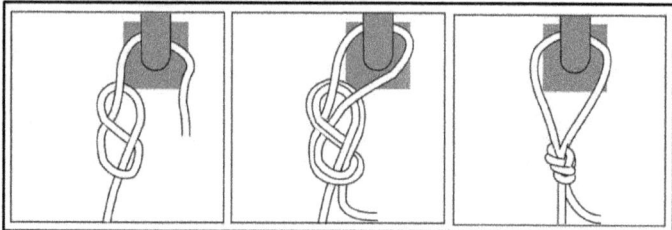

Figure 9-12. Rerouted figure eight knot

9-39. The figure eight slipknot (see figure 9-13) forms an adjustable bight in the middle of a rope. The knot is in the shape of a figure eight. Both ropes of the bight pass through the same loop of the figure eight. The bight is adjustable by means of a sliding section.

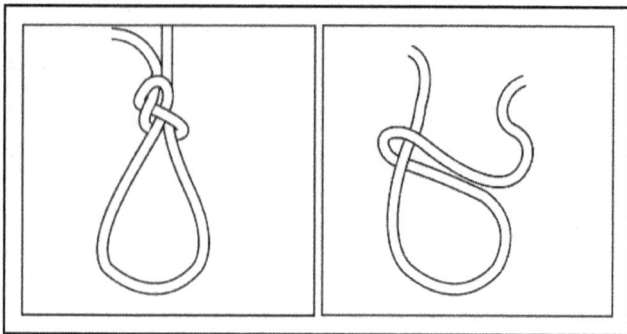

Figure 9-13. Figure eight slipknot

9-40. The <u>Munter hitch</u> (see figure 9-14) is one of the most often used belays and requires very little equipment. Route the rope through a locking, pear-shaped carabiner and then back on itself. The belayer controls the rate of descent by manipulating the working end back on itself with the brake hand.

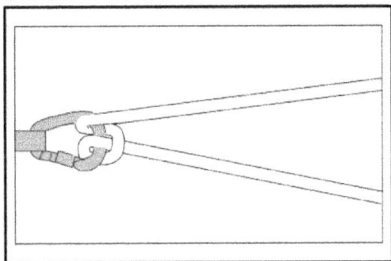

Figure 9-14. Munter hitch

9-41. The <u>Munter mule knot</u> (see figure 9-15) is a knot that allows the user to stop the movement of the rope through a Munter hitch. The Munter mule knot allows an easy release of tension with the pull of a rope and a smooth, controlled release. It is vital knot for many rope systems. Start by tying a Munter hitch in the rope in the loaded position. Maintaining tension, create two loops in the brake strand on either side and behind the main strand. Pass one loop through the other on top of the main strand and carefully pull it tight while maintaining tension. Secure the knot by tying an overhand knot with the bight of rope protruding from the knot. Include the main strand in the knot.

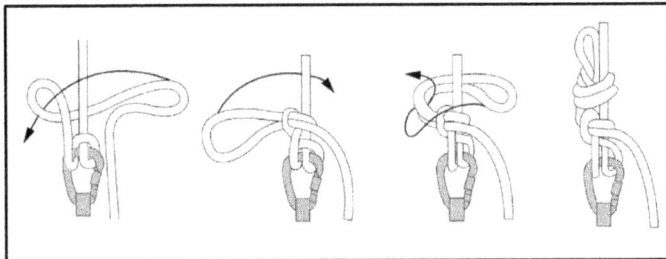

Figure 9-15. Munter mule knot

9-42. A Prusik knot (see figure 9-16) attaches a movable rope to a fixed rope. The knot has two round turns with a locking bar perpendicular to the standing end of the rope. Secure the knot with a double figure eight or bowline within 6 inches (15 centimeters) of the locking bar. When dressing the knot, leave at least a 4-inch (10-centimeter) pigtail on the working end. Tie the Prusik knot in the end or the middle of the rope.

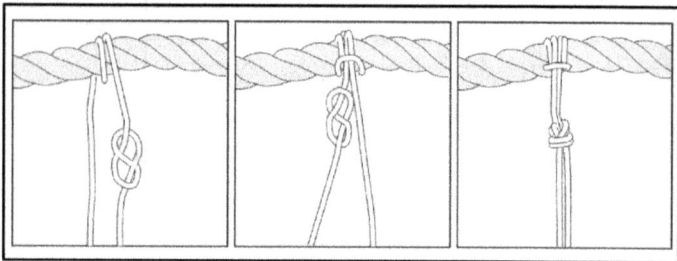

Figure 9-16. Prusik knot

9-43. The <u>bowline</u> (see figure 9-17) is useful for tying the end of a rope around an anchor and for tying a single fixed loop in the end of a rope. Bring the working end of the rope around the anchor from right to left (as the climber faces the anchor). Form an overhand loop in the standing part of the rope (on the climber's right) toward the anchor. Reach through the loop and pull up a bight. Place the working end of the rope (on the climber's left) through the bight and bring it back onto itself. Now, dress the knot down. Form an overhand knot with the tail from the bight.

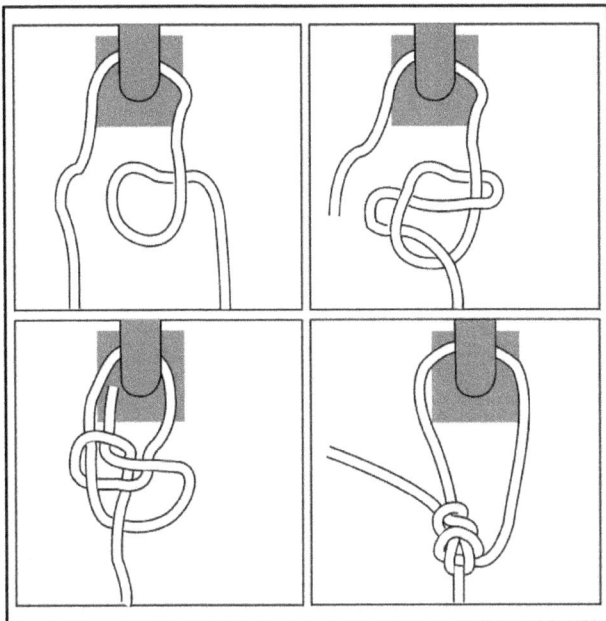

Figure 9-17. Bowline

BELAYS

9-44. <u>Belaying</u> is any action taken to arrest a falling climber or to control the rate of descent of a load from a higher to lower elevation. The belayer also helps manage a climber's rope or the rate of descent of the climber or person rappelling by controlling the amount of rope taken in or out. The belayer is anchored in a stable position to prevent being pulled out of position and losing control of the rope. There are three types of belays: body belays, mechanical belay devices, and friction hitches.

9-45. The <u>body belay</u> (see figure 9-18) uses the belayer's body to apply friction. Routing the rope around the belayer's body creates the friction necessary to arrest a climber's fall. This requires carefulness since the body bears the entire weight of the load.

Figure 9-18. Body belay method

9-46. The <u>mechanical belay</u> uses mechanical devices (see figure 9-19) to help the belayer control the rope such as in rappelling. A variety of mountaineering devices are useful for constructing a mechanical belay. Most mechanical belay devices double as rappel devices.

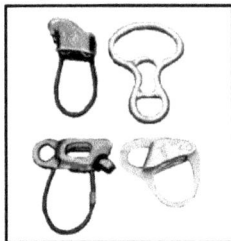

Figure 9-19. Examples of mechanical belay devices

9-47. The <u>air traffic controller</u> is a locking mechanical belay device. (See figure 9-20.) It locks down on itself upon the application of tension in opposite directions. This allows the belayer to apply very little force with the brake hand to control the rate of descent or to arrest a climber's fall.

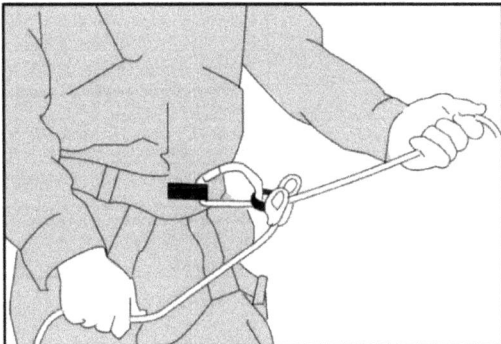

Figure 9-20. Air traffic controller

9-48. Friction hitches such as the Munter hitch (see figure 9-14 on page 9-17) are excellent for belaying loads during the lowering or raising of loads. Table 9-2 demonstrates the sequence of commands that climbers and belayers use.

Table 9-2. Sequence of climbing commands

COMMAND	GIVEN BY	MEANING
BELAY ON, CLIMB	Belayer	Belay is on and climber may climb.
CLIMBING	Climber	Climber is climbing.
UP-ROPE	Climber	Belayer, remove excess slack in rope.
BRAKE	Climber	Belayer, immediately apply brake.
FALLING	Climber	Climber is falling. Belayer, immediately apply brake and prepare to arrest the fall.
TENSION	Climber	Belayer, remove all slack from climbing rope until rope is tight, apply brake, and hold position.
SLACK	Climber	Belayer, allow climber to pull slack into climbing rope and be ready to assist.
ROCK	Anyone	Alert everyone about an object falling near them. Belayer, immediately apply brake.
POINT	Climber	In the event of a fall, alert belayer of change in direction of pull on climbing rope.
STAND BY	Climber or belayer	Hold position and stand by. I am not ready.
DO YOU HAVE ME?	Climber	(Informal) Belayer, prepare for a potential fall, or belayer, lower me.
I HAVE YOU	Belayer	The brake is on, and I am prepared for you (climber) to fall / prepared to lower you.
OFF-BELAY	Climber	Climber is safely in, or it is safe for climber to come off belay.
THREE METERS	Belayer	Alert climber to amount of rope between climber and belayer in either feet or meters.
BELAY OFF	Belayer	I am off belay.

ROPE INSTALLATIONS

9-49. Team construction of rope installations helps units negotiate natural and artificial obstacles. Installation teams comprise a squad-sized element with two to four trained mountaineers. Installation teams deploy early and prepare the AO for safe, rapid movement by constructing various types of mountaineering installations.

9-50. Following the construction of an installation, the squad, or part of it, remains on site to secure and monitor the system, assist with the control of forces across it, and adjust or repair it during use. After the unit passes, the installation team disassembles the system and deploys to another area.

9-51. A fixed rope is anchored in place to help Rangers move over difficult terrain. Its simplest form is a rope tied off on the top of steep terrain. As the terrain becomes steeper or more difficult, fixed rope systems often require intermediate anchors along the route. Planning considerations ask:

- Does the installation allow the mountaineers to bypass the obstacle?
- (Tactical) Can the mountaineers secure the obstacle from construction, through negotiation, to disassembly?
- Is it in a safe and suitable location? Is it easy to negotiate? Does it avoid obstacles?
- Are natural and artificial anchors available?
- Is the area safe from falling rock and ice?

ROPE BRIDGES

9-52. Rope bridges are useful in mountainous terrain for bridging linear obstacles such as streams or rivers whose force of flowing water is too great or whose temperatures are too cold to conduct a wet crossing. Static rope is the proper construction material for rope bridges. The maximum bridge span is half the length of the rope for a dry crossing and three-quarters the length of the rope for a wet crossing. An anchor knot on the far side of the obstacle anchors the ropes, and a transport-tightening system ties off the ropes at the near end. (See figure 9-21 on page 9-24.) Rope bridge planning considerations ask:

- Does the installation allow the mountaineers to bypass the obstacle?
- (Tactical) Can the mountaineers secure the obstacle from construction, through negotiation, to disassembly?
- Is it in the most suitable location such as a bend in the river? Is it easily secured?
- Does it have near side and far side anchors?
- Does it have good loading and off-loading platforms?
- Do the mountaineers have the necessary equipment for a one-rope bridge? This includes—
  - One sling rope for every Ranger.
  - One locking, steel carabiner.
  - Three oval-shaped, steel carabiners.
  - Two 120-foot static ropes.

Bridge Construction

9-53. The first Ranger swims the rope to the far side and ties a tensionless anchor, between knee and chest levels, with a minimum of four wraps. The bridge team commander (BTC) ties a transport-tightening system (see figure 9-21 on page 9-24) to the near side anchor point. To do this, tie a figure eight slipknot and incorporate a locking half hitch around the adjustable bight. Insert two oval-shaped, steel carabiners into the bight so the gates are opposite and opposed. Then, route the rope around the near side anchor point at waist level and drop it into the oval-shaped, steel carabiners.

9-54. A three-person pull team moves forward from the platoon. No more than three tighten the rope. More can overtighten the rope, bringing it near failure. Once the rope bridge is tight enough, the bridge team secures the transport-tightening system (see figure 9-21) using two half hitches without losing more than 4 inches (10 centimeters) of tension. Personnel cross, using the commando crawl (see figure 9-22), Tyrolean traverse (see figure 9-23), or monkey crawl (see figure 9-24 on page 9-26) methods.

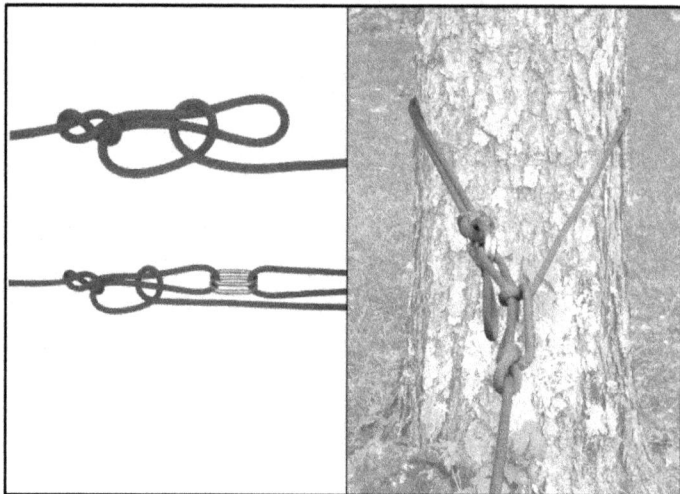

Figure 9-21. Example of transport-tightening system

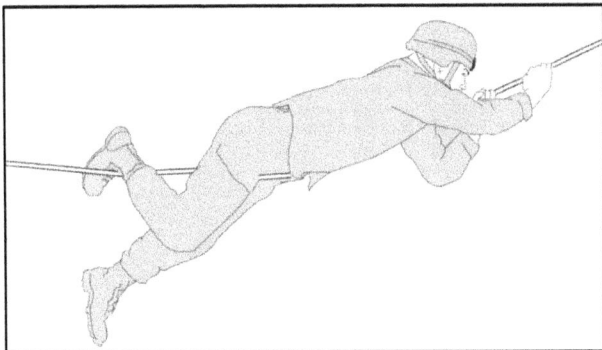

Figure 9-22. Commando crawl method

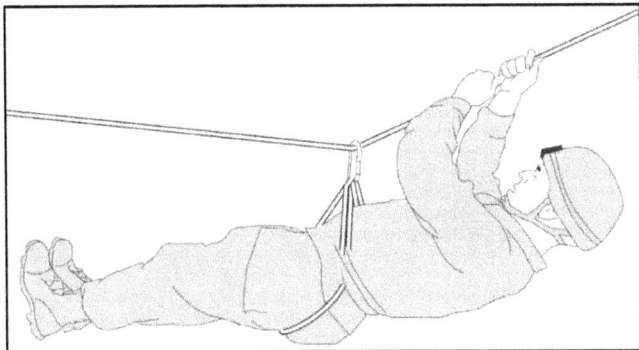

Figure 9-23. Rappel seat (Tyrolean traverse) method

Figure 9-24. Monkey crawl method

Bridge Recovery

9-55. Once all but two troops have crossed the bridge, the BTC chooses the wet or dry method to dismantle the bridge. For the dry method, anchor the tightening system with the transport knot. For a wet crossing, any method is effective for anchoring the tightening system. The procedure for bridge recovery follows.

    a. The BTC back-stacks all the slack coming out of the transport knot, ties a fixed loop, and places a carabiner into the fixed loop.

    b. The next-to-last Ranger to cross attaches carabiner to rappel seat or harness and then moves across the bridge using the Tyrolean traverse method (see figure 9-23 on page 9-25).

    c. The BTC removes all the knots from the system. The far side remains anchored. The rope should now pass only around the near side anchor.

    d. A three-Ranger pull team, assembled on the far side, takes the end brought across by the next-to-last Ranger, pulls, and holds the rope tight again.

    e. The BTC attaches him- or herself to the rope bridge and moves across.

    f. Once across, the BTC breaks down the far side anchor, removes the knots, and pulls the rope across.

        (1) During a wet crossing, all personnel cross except the BTC or the strongest swimmer.

        (2) The BTC then removes all the knots from the system.

        (3) The BTC ties a fixed loop, inserts a carabiner, attaches it to rappel seat or harness, and manages the rope as the far side pulls the slack.

        (4) The BTC then moves across the obstacle under the safety of a belay from the far side.

Z-pulley System

9-56. The Z-pulley system (see figure 9-25 on page 9-28) is a simple, easily constructed hauling system. Anchors must be sturdy and able to support the weight of the load. Different factors such as the tactical situation, weather, terrain, equipment, load weight, and availability of anchors govern site selection. Carabiners substitute for pulleys when the latter are unavailable. The leverage obtained using a Z-pulley system is a three-to-one mechanical advantage. The less friction involved, the greater the mechanical advantage. The rope running through the carabiners, the load rubbing against the rock wall, and the rope's condition all cause friction. The procedure to construct a Z-pulley system follows.

    a. Establish an anchor Prusik system.

    b. Place a carabiner on the runner at the anchor point, place a pulley into the carabiner, and run the hauling rope through the pulley.

    c. With a sling rope, tie a middle-of-the-rope Prusik knot secured with a figure eight knot on the load side of the pulley. This serves as a progress capture device. Alternatively, a mechanical descender serves well in place of the Prusik knot.

    d. Take the tails exiting the figure eight and tie a Munter hitch secured by a mule knot. Ensure the Munter hitch is loaded properly before tying the mule knot.

    e. At an angle away from the anchor Prusik system, establish a moveable Prusik system to create a Z in the hauling rope.

    f. Tie another Prusik knot on the load side of the hauling rope secured with a figure eight knot. Using the tails, tie a double-double figure eight knot.

    g. Insert a locking carabiner into the two loops formed and then place the working end into the carabiner. Mechanical ascenders are not to serve as a moveable Prusik system.

    h. Move the working end back on a parallel axis with the anchor Prusik system. Provide a pulling team on the working end with extra personnel to monitor the Prusik knots.

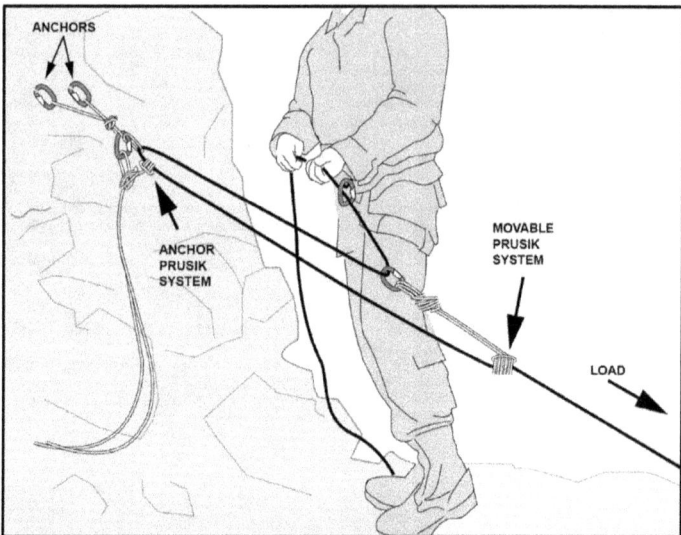

Figure 9-25. Z-pulley system

RAPPELLING

9-57. Rappelling is a quick method of descent, but it can be extremely dangerous. Dangers include human error and the failure of the anchor or other equipment. Carefully choose anchors for use in mountainous environments. Take great care to load the anchor slowly. Also, avoid placing too much stress on the anchor. The standing prohibition on bounding rappels enforces these safety practices, only walk down rappels are permissible.

9-58. Body rappels are quick and easy (see figure 9-26) but for use on moderate pitches only—never on vertical or overhanging terrain. The use of gloves prevents rope burns.

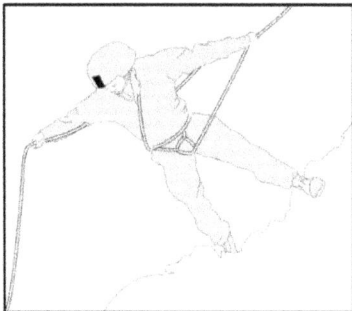

Figure 9-26. Body rappel method

9-59. Seat hip rappels use either a figure eight descender (see figure 9-27) or a carabiner wrap descender (see figure 9-28 on page 9-30). Insert the descender into a sling rope seat and then fasten it to the person rappelling. This gives the person enough friction for a smooth, controlled descent.

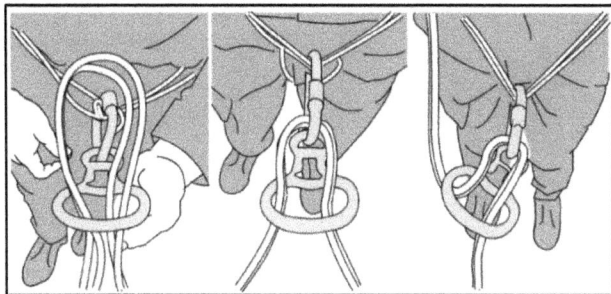

Figure 9-27. Figure eight descender

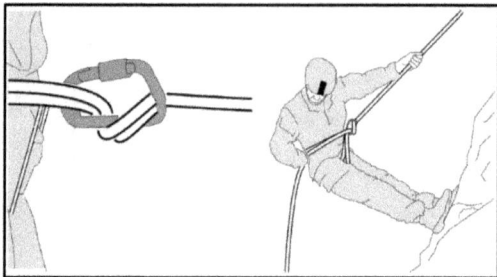

Figure 9-28. Closeup of carabiner wrap descender and the seat hip rappel (shown with carabiner wrap seat hip descender)

9-60. For an <u>extended air traffic controller rappel</u>, the controller is to extend away from the body when possible. A sling rope is useful for accomplishing this. (See figure 9-29.) This lowers the center of gravity of the person rappelling, rendering the edge transition smoother. The use of both hands below the device controls the descent. (See figure 9-30.)

Figure 9-29. Sling extension

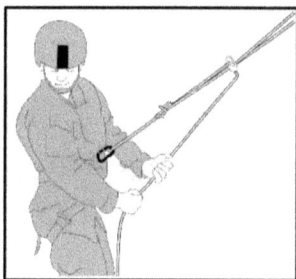

Figure 9-30. Extended air traffic controller rappel

Chapter 10

Machine Gun Employment

Machine guns are a Ranger platoon's most effective weapon against the dismounted enemy force. Machine guns allow the Ranger unit to engage enemy forces from a greater range and with greater accuracy than individual weapons.

SPECIFICATIONS

10-1. A leader's abilities to employ available machine guns properly and achieve fire superiority are often the deciding factors in the battlespace. Table 10-1 provides references and specifications for various machine guns. Table 10-2 on pages 10-2 and 10-3, figure 10-1 on page 10-3, and figure 10-2 on page 10-4 provide definitions associated with machine guns and some of their depictions.

Table 10-1. Specifications of machine guns

WEAPON	M249	M240B	M2	MK19
Information	TC 3-22.249 TM 9-1005-201-10	TC 3-22.240 TM 9-1005-313-10	TC 3-22.50 TM 9-1005-213-10	TC 3-22.19 TM 9-1010-230-10
Description	5.56-mm gas-operated automatic	7.62-mm gas-operated medium	Caliber .50 recoil-operated heavy	40-mm air-cooled, blowback-operated automatic GL
Weight	Gun with barrel – 16.41 pounds (7.4 kilograms), Tripod – 16 pounds (7.3 kilograms)	Gun with barrel – 27.6 pounds (12.5 kilograms), Tripod – 20 pounds (9.1 kilograms)	Gun with barrel and tripod – 128 pounds (58 kilograms)	Gun with barrel and tripod – 140.6 pounds (63.8 kilograms)
Length	104 centimeters (41 inches)	110.5 centimeters (43.5 inches)	156 centimeters (61 inches)	109.5 centimeters (43 inches)
Max range	3,600 m	3,725 m	6,764 m	2,212 m
Max effective range	Bipod/point: 600 m Bipod/area: 800 m Tripod/point: 800 m Tripod/area: 1,000 m Grazing: 600 m	Bipod/point: 600 m Bipod/area: 800 m Tripod/point: 800 m Tripod/area: 1,100 m Grazing: 600 m Suppression: 1,800 m	Point: 1,500 m (single shot) Area: 1,830 m Grazing: 700 m	Point: 1,500 m Area: 2,212 m
Tracer BO	900 m	900 m	1,800 m	None

Table 10-1. Specifications of machine guns (continued)

WEAPON	M249	M240B	M2	MK19
Sustained rate of fire	50 rpm 6 to 9 rounds 4 to 5 seconds Every 10 minutes	100 rpm 6 to 9 rounds 4 to 5 seconds Every 10 minutes	40 rpm 6 to 9 rounds 10 to 15 seconds End of day or when damaged	40 rpm
Rapid rate of fire	100 rpm 6 to 9 rounds 2 to 3 seconds 2 minutes	200 rpm 10 to 13 rounds 2 to 3 seconds 2 minutes	40 rpm 6 to 9 rounds 5 to 10 seconds End of day or when damaged	60 rpm
Cyclic rate of fire	850 rpm. continuous burst	650 to 950 rpm, continuous burst	450 to 550 rpm, continuous burst	325 to 375 rpm, continuous burst

Legend: BO—burnout;   GL—grenade launcher;   m—meter;   max—maximum;   mm—millimeter; rpm—revolutions per minute; TC—training circular; TM—technical manual

Table 10-2. Machine gun terms

Line of sight	The imaginary line drawn from the firer's eye through the sight to the point of aim.
Burst of fire	A number of successive rounds fired with the same elevation and point of aim when the trigger is held to the rear. The number of rounds in a burst varies depending on the type of fire employed.
Trajectory	The curved path of the projectile in its flight from the muzzle of the weapon to its impact. As the range to the target increases, so does the curve of trajectory.
Maximum ordinate	The height of the highest point above the line of sight the trajectory reaches between the muzzle of the weapon and the base of the target. It always occurs at a point about two-thirds of the distance from the weapon to the target and increases with range.
Cone of fire	The pattern formed by the different trajectories in each burst as they travel downrange. The vibration of the weapon and variations in ammunition and atmospheric conditions all contribute to the trajectories that compose the cone of fire.
Beaten zone	The elliptical pattern formed when the rounds in the cone of fire strike the ground or target. Though normally oval- or cigar-shaped, the size and shape of the beaten zone changes as a function of the range to target and slope of the target and as the density of the rounds decreases toward the edges. Gunners and automatic rifle shooters engage targets to take maximum effect of the beaten zone. Due to the right-hand twist of the barrel, the simplest way to do this is to aim at the left base of the target.

Table 10-2. Machine gun terms (continued)

Sector of fire	An area to be covered by fire and assigned to an individual, a weapon, or a unit. Leaders normally assign a primary and a secondary sector of fire to gunners.
Primary sector of fire	An assignment to the gun team to cover the most likely avenue of enemy approach from all types of defensive positions.
Secondary sector of fire	An assignment to the gun team to cover the second most likely avenue of enemy approach. Fired from the same gun position as the primary sector of fire.
Final protective fire (FPF)	An immediately available, prearranged barrier of fire to stop enemy movement across defensive lines or areas.
Final protective line (FPL)	A predetermined line of grazing fire to stop an enemy assault. Sight the machine gun along an assigned FPL except when the unit is engaging other targets. An FPL becomes part of the unit's machine gun FPFs. An FPL is fixed in direction and elevation. However, a unit employs a small shift for search to prevent the enemy from crawling under the FPL and to compensate for irregularities in the terrain or the sinking of the tripod legs into soft soil during firing. Gunners deliver fire during all conditions of visibility.
Principal direction of fire (PDF)	An assignment to a gunner to cover an area with either good fields of fire or a likely dismounted avenue of approach. A PDF provides mutual support to an adjacent unit. With no assigned FPL, sight machine guns use the PDF. With an assigned PDF and when the unit is not engaging other targets, machine guns remain on the PDF. A leader uses a PDF only when not assigning an FPL; it then becomes the machine gun's part of the unit's FPFs.

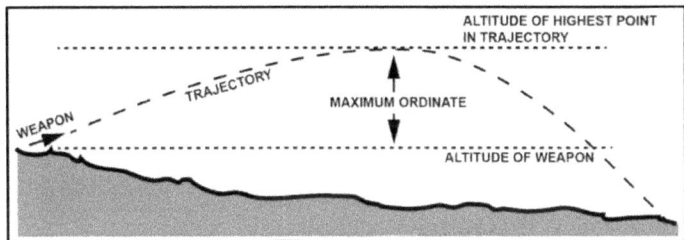

Figure 10-1. Trajectory and maximum ordinate

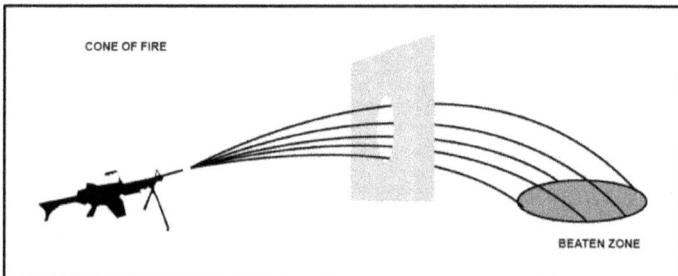

Figure 10-2. Cone of fire and beaten zone

CLASSES OF AUTOMATIC WEAPON FIRE

10-2. The U.S. Army classifies automatic weapon fire with respect to the ground, target, and weapon. Table 10-3 and figure 10-3 detail classes of fire with respect to the ground.

Table 10-3. Classes of fire with respect to ground

Grazing	The center of the cone of fire rises less than 1 meter (3 feet) above the ground. Employed in the final protective line defense. Possible only with level or uniformly sloping terrain. Indirect fire such as that from an M320 covers any dead space encountered along the final protective line. When firing over level or uniformly sloping terrain, the M240B and M249 attain a maximum of 600 meters of grazing fire. The M2 attains a maximum of 700 meters.
Plunging	The danger space is inside the beaten zone. Weapons fire at long range, from high to low ground, into abruptly rising ground or across uneven terrain resulting in a loss of grazing fire at any point along the trajectory.

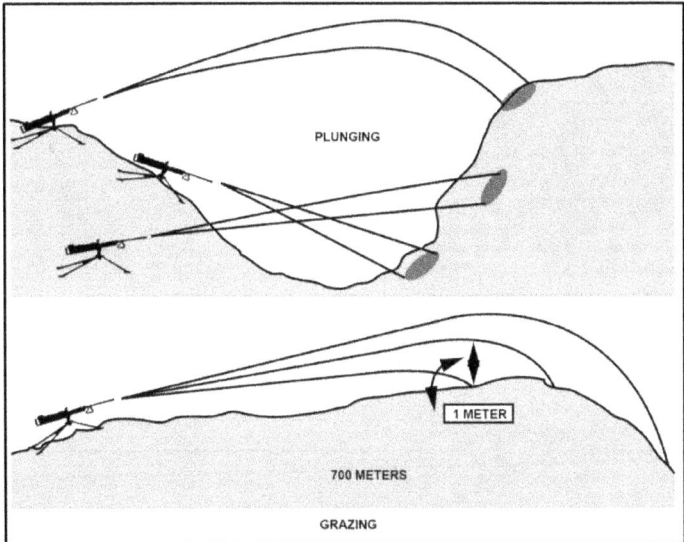

Figure 10-3. Classes of fire with respect to ground

10-3. Leaders and gunners strive at all times to position their gun teams where they can best take advantage of the machine gun's beaten zone with respect to an enemy target. Channeling the enemy by using terrain or obstacles so they approach a friendly machine gun position from the front in a column formation is one example.

10-4 In this situation, the machine gun employs enfilade fire on the enemy column, and the effects of the machine gun's beaten zone are much greater than if it had engaged the same enemy column from the flank. Table 10-4 defines and compares the four classes of fire with respect to the target. Figure 10-4A and figure 10-4B on page 10-8 depict these classes.

Table 10-4. Classes of fire with respect to target

Enfilade (best)	Occurs when long axes of beaten zone and target coincide or nearly coincide.
	Can be frontal fire on a column or flanking fire on a line.
	Most desirable class of fire with respect to the target:
	    Makes maximum use of the beaten zone.
	    Leaders and gunners always try to position guns for enfilade fire.

Frontal (Column: Yes, Line: No)	Occurs when the long axis of the beaten zone is at a right angle to the front of the target. Highly desirable against a column. Becomes enfilade fire as the beaten zone coincides with the long axis of the target. Less desirable against a line because most of the beaten zone normally falls below or after the enemy target.
Flanking (Column: No, Line: Yes)	Delivered directly against the flank of the target. Most desirable against a line. Becomes enfilade fire as the beaten zone coincides with the long axis of the target. Least desirable against a column because most of the beaten zone normally falls before or after the enemy target.
Oblique	For gunners and automatic rifle shooters. The long axis of the beaten zone is at any angle other than a right angle to the front of the target.

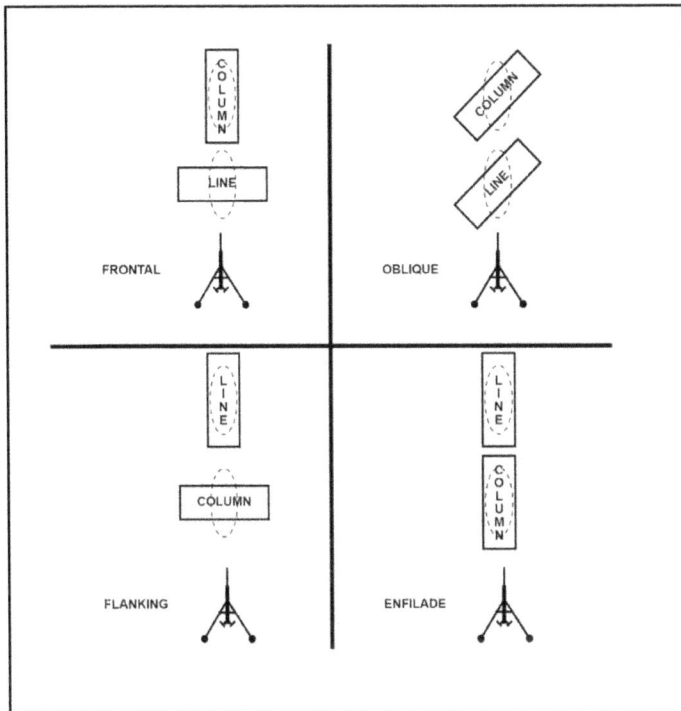

Figure 10-4A. Classes of fire with respect to target

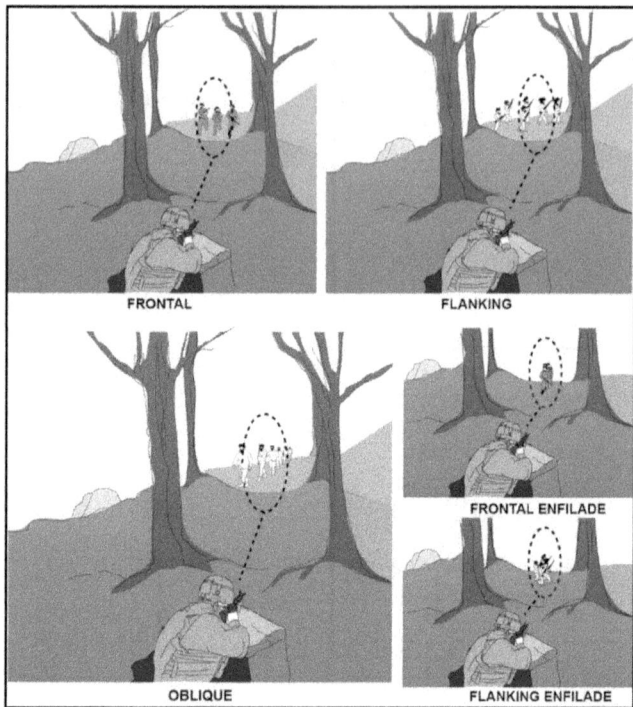

Figure 10-4B. Classes of fire with respect to target

10-5. Fires with respect to the machine gun include fixed, traverse, search, traverse and search, swinging traverse, and free gun. Table 10-5 describes these classes, and figure 10-5 on page 10-10 depicts them.

Table 10-5. Classes of fire with respect to gun

Fixed	Delivered against a stationary point target when the depth and width of the beaten zone cover the target with little or no manipulation. After the initial burst, the gunners follow any change in or movement of the target without command.
Traverse	Disperses fires in width by successive changes in direction but not elevation. Delivered against a wide target with minimal depth. When engaging a wide target requiring traverse fire, the gunner selects successive aiming points throughout the target area. These aiming points are close enough together to ensure adequate target coverage. However, they need not be so close that they waste ammunition by concentrating a heavy volume of fire in a small area.
Search	Distributes fires in depth by successive changes in elevation. Employed against a deep target or a target with depth and minimal width requiring changes only in the elevation of the gun. The amount of elevation change depends on the range and slope of the ground.
Traverse and search	A combination in which successive changes in direction and elevation result in the distribution of fire in width and depth. Employed against a target whose long axis is oblique to the direction of fire.
Swinging traverse	Employed against targets requiring major changes in direction but little or no change in elevation. Targets can be dense, wide, in close formations moving slowly toward or away from the gun, or vehicles or mounted troops moving across the front. When tripod-mounted, the traversing slide lock lever is loosened enough to permit the gunner to swing the gun laterally. With swinging traverse fire, the weapon normally fires at the cyclic rate of fire. Consumes a lot of ammunition and does not have a beaten zone because each round seeks its own area of impact.
Free gun	Delivered against rapidly engaged moving targets with fast changes in direction and elevation (for example, aerial targets, vehicles, mounted troops, enemy soldiers in relatively close formations moving rapidly toward or away from the gun position). With free gun fire, the weapon normally fires at the cyclic rate of fire. Consumes a lot of ammunition and does not have a beaten zone because each round seeks its own area of impact.

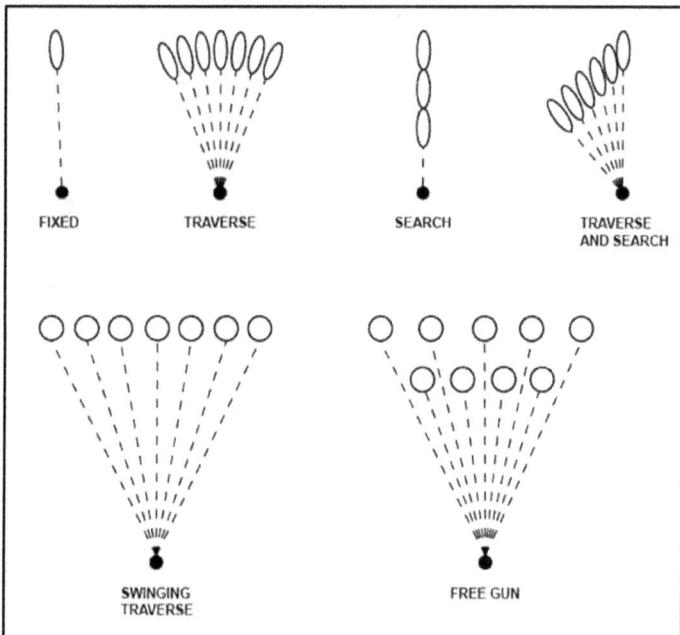

Figure 10-5. Classes of fire with respect to gun

OFFENSE

10-6. Successful offensive operations depend on the effective employment of fire and movement. They are essential and dependent upon each other. For example, without the support of covering fire, maneuvering in the presence of enemy fire can produce huge losses.

10-7. Covering fire, especially that which provides fire superiority, allows maneuvering in the offense. However, fire superiority alone rarely wins battles. The primary objective of the offense is to advance on, occupy, and hold the enemy position.

MEDIUM MACHINE GUNS

10-8. In the offensive, the PL can establish a base-of-fire element with the M240B, the M249 light machine gun, or a combination of the weapons. When the platoon scheme of maneuver is to conduct the assault with the Infantry squads, the PSG or WSL may position this element and control its fires. The M240B machine gun is more stable and accurate at greater ranges but takes longer to maneuver on the tripod than on the bipod. Machine gunner responsibilities are to—

- Target key enemy weapons until the enemy's assault element masks the machine gunners' fires.
- Suppress the enemy's ability to return accurate fire.
- Hamper the maneuver of the enemy's assault element.
- Fix the enemy in position.
- Isolate the enemy by cutting off their avenues of reinforcement.
- Shift fires to the flank opposite the one being assaulted and continue targeting any automatic weapons providing enemy support.
- Engage any enemy counterattack.
- Cover the gap created between the forward element of the friendly assaulting force and the terrain covered by indirect fire when direct fire lifts and shifts.
- On signal, displace (with the base-of-fire element) to join the assault element on the objective.

MK19 AND M2

10-9. As part of the base-of-fire element, the MK19 and M2 help the friendly assault element. The base-of-fire element does this by suppressing enemy bunkers and lightly armored vehicles. Even when their fire is too light to destroy enemy vehicles, well-aimed suppressive fire keeps the enemy buttoned up and unable to place effective fire on friendly assault elements.

10-10. The MK19 and M2 are particularly effective in preventing lightly armored enemy vehicles from escaping or reinforcing. Both vehicle-mounted weapons fire from a long-range standoff position or travel forward with the assault element.

BASE OF FIRE

10-11. Machine gun fire from a support by fire position is the necessary minimum for keeping the enemy from returning effective fire. Gunners conserve ammunition. The WSL positions and controls the fires of all machine guns in the element.

10-12. Machine gun targets include key enemy weapons or groups of enemy targets either on the objective or attempting to reinforce or counterattack. The nature of the terrain, the desire to achieve some standoff, and the other METT-TC (I) variables prompt the leader to the correct tactical positioning of the base-of-fire element. There are distinct phases of rates of fire that the base-of-fire element employs:

- Initial heavy volume (rapid rate) to gain fire superiority.
- Slower (sustained) rate to conserve ammunition while preventing effective return fire as the assault moves forward.
- Increased rate as the assault nears the objective.
- Lift and shift to targets of opportunity.

10-13. Machine guns in the support by fire role are set in role and assigned primary and alternate sectors of fire and primary and alternate positions. Machine guns are suppressive fire weapons for known and suspected enemy positions. Therefore, gunners do not empty all their ammunition into one bunker simply because that is all they have identified at the time.

10-14. Shift and shut down the weapons squad gun teams one at a time, not all at once. M203, mortar, and other indirect fire are useful for suppression during the relocation of machine guns. Leaders take into account the surface danger zone of the machine guns when planning and executing the lift or shift of the support by fire guns. The effectiveness of the enemy on the objective plays a large role in how much risk leaders take with respect to the lifting or shifting of fires. Once the support by fire line is masked by the assault element, gunners lift or shift fires, or lift and shift, to prevent enemy withdrawal or reinforcement.

MANEUVER ELEMENT

10-15. Under certain terrain conditions and for proper control, machine guns join the maneuver or assault unit. When this is the case, leaders assign them a cover fire zone or sector. The machine guns seldom accompany the maneuver element. The gun's primary mission is to provide covering fire. The maneuver element employs machine guns only when the area or zone of action assigned to the assault unit or company is too narrow to permit proper control of the guns. The unit then moves the machine guns and readies them to employ on order from the leader and in the direction requiring supporting fire.

10-16. When the machine guns are to move with the element undertaking the assault, the maneuver element brings the machine guns to provide additional firepower. Gunners fire these weapons from bipods, in an assault mode, from the hip, or from the underarm position. They target enemy automatic weapons anywhere on the unit's objective.

10-17. After destroying any enemy automatic weapons, the gunners distribute fires over their assigned zones or sectors. The machine gunner in the assault position engages within 300 meters of the target often at point-blank ranges.

10-18. When the platoon's organic weapons fail to cover the area or zone of action, the company commander has the option of assigning more machine guns and personnel to help the platoon accomplish its assigned mission. Leaders assign each machine gunner a zone or a sector to cover, and they move with the maneuver element.

CONTROLLED OCCUPATION AND WITHDRAWAL OF THE SUPPORT BY FIRE POSITION

10-19. Controlled occupation of the support by fire position is one of the key elements in setting up a support by fire position. To remain undetected, use stealth and control. Rangers follow these procedural steps to occupy their support by fire position effectively.

    a. The WSL moves to and establishes an RP just short of the support by fire position.

Note. The order of movement for the weapons squad during movement to their position is the WSL, gun team 1 (gunner, ammunition bearer, and assistant gunner), gun team 2 (assistant gunner, gunner, ammunition bearer), and gun team 3 (assistant gunner, gunner, ammunition bearer).

---

    b. The WSL then moves forward from the RP with the gun 1 gunner, followed by the gun 1 ammunition bearer to link ammunition, and finally the gun 1 assistant gunner.

      (1) The gunner gets into position and remains in bipod mode to provide security.

      (2) Once gun 1 is on the bipod with ammunition linked and the spare barrel out, the WSL notifies the PL that the support by fire is <u>occupied</u>.

    c. The WSL next brings forward the gun 2 assistant gunner, gunner, and ammunition bearer.

      (1) The assistant gunner sets the tripod, and the gunner sets gun 2 on the tripod.

      (2) The ammunition bearer drops off all ammunition at the gun position, links the ammunition, prepares the spare barrel, and moves to pull flank or rear security.

    d. Once gun 2 is in place, the WSL brings forward the gun 3 assistant gunner, gunner, and ammunition bearer.

      (1) The assistant gunner sets the tripod, and the gunner sets gun 3 on the tripod.

      (2) The ammunition bearer drops off all ammunition at the gun position, links the ammunition, prepares the spare barrel, and moves to pull flank or rear security.

    e. Upon the emplacement of gun team 3, gun team 1 moves the gun from bipod to tripod.

    f. Upon the emplacement of the support by fire, the WSL gets down behind the guns to ensure they cover their sectors of fire and that everything is in accordance with the PL's guidance.

    g. The WSL calls the PL and reports the support by fire position as <u>established</u>.

10-20. The PL sometimes uses controlled withdrawal of the support by fire position to cover the withdrawal of the platoon and provide security for the support by fire position. Rangers follow these procedural steps to withdraw from their support by fire position.

    a. Before the platoon moves off the objective, the WSL shifts the machine guns' sectors of fire to cover the objective from enemy reinforcements.

    b. After the guns are covering the objective, the WSL starts breaking down the gun positions one at a time.

    c. After the main body of the platoon begins moving off the objective, the gun teams move one at a time into the order of movement with the last gun breaking down as soon as the platoon is completely off the objective.

    d. The entire weapons squad moves tactically to link up with the rest of the platoon.

DEFENSE

10-21. The platoon's defense centers on its machine guns. The PL positions the rifle squad to protect the machine guns against the assault of a dismounted enemy formation.

10-22. The machine gun provides the necessary range and volume of fire to cover the squad's front in the defense. However, position is very important. The requirements to employ and position machine guns follow:

- The main requirement of a suitable machine gun position in the defense is its effectiveness in accomplishing specific missions.
- The best positions are accessible and afford cover and concealment. Machine guns in such positions protect the front, flanks, and rear of occupied portions of the defensive position and are mutually supportive.
- Attacking troops usually seek easily traveled ground that provides cover from fire.
- For each machine gun, the leader chooses three positions: primary, alternate, and supplementary. The redundancy ensures they cover the sector and have protection on their flanks.
- The leader positions each machine gun to cover the entire sector or to overlap sectors with the other machine guns.
- The engagement range extends from over 1,000 meters, where the enemy begins its assault, to point-blank range.
- Machine gun targets include enemy automatic weapons and command and control elements.

10-23. Defensive positions distribute machine gun fire in width and depth. Machine guns are the backbone or framework of the defense. Firstly, leaders use them to subject the enemy to increasingly devastating fire from the initial phases of the attack. Secondly, leaders use them to neutralize any partial enemy successes by delivering intense fires in support of counterattacks. Lastly, their tremendous firepower helps the unit hold ground.

10-24. In the defensive, the <u>medium machine gun</u> provides sustained direct fire that covers the most likely or most dangerous enemy dismounted AAs. It also protects friendly units against the enemy's dismounted close assault. The PL positions the machine guns to concentrate fires in locations with the capacity to inflict the most damage to the enemy. The leader also places them to take advantage of grazing enfilade fire, standoff or maximum engagement range, and the best observation of the target area.

10-25. Machine guns provide overlapping and interlocking fires with adjacent units and cover tactical and protective obstacles with traverse or search fire. When leaders call for final protective fire, machine guns (aided by M249 fire) place an effective barrier of fixed direct fire across the platoon's front. Leaders position machine guns to—

- Concentrate fires where they want to kill the enemy.
- Fire across the platoon's front.
- Cover obstacles by direct fire.
- Tie in with adjacent units.

10-26. In the defensive, the MK19 and M2 machine guns fire from the vehicle mount or dismounted from the vehicle and mounted on a tripod at a defensive fighting position appropriate to the weapon system. These guns provide sustained direct fire that covers the most likely enemy mounted AA. Their maximum effective range enables them to engage enemy vehicles and equipment at far greater ranges than the platoon's other direct-fire weapons.

10-27. When mounted on the tripod, the M2 and MK19 are highly accurate to their maximum effective ranges. They allow leaders to plan predetermined fires for likely high-payoff targets. The tradeoff is these weapon systems are heavy, and a unit's movement of them is slow. These guns are less accurate when mounted on vehicles than when fired from the tripod mount system. However, units more easily maneuver them to alternate firing locations whenever the need arises.

CONTROL OF MACHINE GUNS

10-28. Leaders use control measures, coordinating instructions, and fire commands to control the engagements of their machine guns. Rehearsals are key in a leader's ability to control machine guns. The noise and confusion of battle can limit the ability of a leader to control the machine guns. Therefore, a leader uses a combination of methods that accomplish the mission. Several successful methods for a leader to control fires are oral, hand and arm signals, prearranged signals, personal contact, and range cards.

10-29. A leader gives a fire command to deliver effective fire on a target quickly and without confusion. It is essential that troops understand the WSL's delivered commands and that the assistant gunner or gun TL and the gunner echo them. The elements of a fire command operate as follows:

- Alert lets the gun team know they are about to engage a target.
- Direction lets the gun team know where to engage.
- Description lets the gun team know what they are engaging.
- Range (if not already set on a predestined target) allows the gun team to adjust the traverse and elevation mechanism.
- Method of fire includes manipulation and rate of fire.
- Manipulation dictates the class of fire with respect to the weapon by the announcement of FIXED, TRAVERSE, SEARCH, or TRAVERSE AND SEARCH.
- Rate controls the volume of fire (sustained, rapid, and cyclic).
- Command to open fire initiates the firing of the weapon system.

10-30. Leaders carefully plan the machine guns' rates of fire appropriate for the mission and the amount of ammunition available. The WSL fully understands the mission, the amount of available ammunition, and the application of necessary machine gun fire to support all key events of the mission fully. Careful planning helps ensure the guns do not run out of ammunition.

10-31. A mounted platoon has a chance of having access to enough machine gun ammunition to support the guns throughout an operation. On the other hand, a dismounted platoon with a limited resupply capability plans on having available only the basic load. In either case, leaders take into account the mission's key events that the guns support. They plan for the necessary rate of machine gun fire to support the key events and the requisite amount of ammunition for the scheduled rates of fire.

10-32. The leader estimates how much ammunition is necessary to support all the machine guns and adjusts the amount each event uses to ensure enough ammunition remains for all phases of the operation. Examples of planning rates of fire and ammunition requirements for a platoon's machine guns in the attack follow.

Weapons Squad Tactics, Techniques, and Procedures

A. Use a starter belt (about 50 to 70 rounds) when moving.

B. Ensure ammunition and NVDs are readily accessible and in packs such as an assault pack for mounted and city operations or a rucksack for long sustainment missions.

C. Carry the traverse and elevation mechanism and the tripod together.

D. When the tripod is taken is mission dependent such as in urban operations.

E. Use optics, lasers, NVDs. For example, in urban operations, consider using a reflexive sight because most engagements are 150 meters or closer. Also, zero the iron sights.

Chapter 11

Urban Operations

Today's security environment demands more from leaders than ever before. Leaders do not only lead Rangers but also influence other people. They are able to work with other members of the armed services and government agencies. They win the willing cooperation of multinational partners, both military and civilian. Urban offensive operations pose great risks to Army forces and noncombatants. Yet, the military demands self-aware and adaptive leaders who compel enemies to surrender in war and master the circumstances facing them in stability operations and peace. Victory and success depend on the effectiveness of these leaders' organizations. Developing effective organizations requires hard, realistic, and relevant training.

PLANNING

11-1. Urban operations include decisive action—continuous combinations of offensive, defensive, and stability operations or defense support of civil authorities' tasks. An urban operation executes these tasks either sequentially or (more likely) simultaneously. (For further study of urban operations, see ADP 3-0, ATP 3-06.11, and ATP 3-21.8.)

11-2. Urban areas are strategically important. Several factors attract armies to combat in urban areas. Some common reasons follow:
- Using the defensive advantages of the urban environment.
- Developing allegiance and support from among the populace.
- Adapting urban resources such as the infrastructure or capabilities for operational or strategic purposes.
- Drawing in the enemy.
- Playing on an area's symbolic importance.
- Using an area's geographical advantages, especially to develop an AA or to dominate a region.

11-3. Task-organizing subordinate units for urban operations depends largely on the nature of the operation. Some units are always part of the task organization to ensure the success of urban operations. Infantry, special operations, civil affairs, aviation, military police, military information support operations, military intelligence, and engineers are units required for decisive action in urban operations. Other types of forces such as Armor, artillery, and chemical units have essential roles in specific types of urban operations but are less applicable to other operations.

11-4. Military forces conduct decisive action in urban areas. Commanders conduct decisive action abroad by executing offensive, defensive, and stability urban operations as part of a joint, interagency, and multinational effort. The situation mandates the dominion of one type of operation—offense, defense, stability, or defense support of civil authorities—over the urban operation. Commanders often find themselves executing offensive, defensive, and stability operations at the same time. In fact, waiting for the conclusion of all combat operations before beginning stability operations often results in lost, sometimes irretrievable, opportunities. The dominant type of operation varies between different urban areas, even in the same campaign.

PREPARATION

11-5. Operating successfully in a complex urban environment requires a thorough understanding of the environment and rigorous, realistic urban operations training. Training covers every aspect of decisive action including appropriate tactics, techniques, and procedures related to offense, defense, and stability operations. Training also replicates the following conditions:

- The psychological impact of intense, close combat against a well-trained, relentless, and adaptive enemy.
- The effects of noncombatants including government and nongovernmental organizations and agencies in close proximity to Army forces. These necessitate—
  - An in-depth understanding of culture and its effects on perceptions.
  - An understanding of civil administration and governance.
  - The ability to mediate and negotiate with civilians including effective communication through an interpreter.
  - The development and use of flexible, effective, and understandable rules of engagement (ROE).
- A complex intelligence environment requiring lower echelon units to collect and forward essential information to higher echelons for rapid synthesis into timely and useable intelligence at all levels of command. The multifaceted urban environment requires a bottom-fed approach to developing intelligence.
- The communications challenges imposed by the environment, as well as the need to transmit large volumes of information and data.
- The medical and logistical problems associated with operations in an urban area including constant threat interdiction against lines of communication and sustainment bases.

11-6. In a complex urban environment, every Ranger, regardless of branch or military occupational specialty, remains committed and prepared to close with and kill or capture threat forces. Every Ranger also remains prepared for effective interaction with the urban area's noncombatant population and to assist in the unit's information collection efforts.

11-7. In urban operations, every Ranger is likely to perform advanced rifle marksmanship including advanced firing positions, short-range marksmanship, and night firing techniques (unassisted and with the use of optics). While not all-inclusive or necessarily urban-specific, other critical individual and collective urban operations tasks include those to—

- Conduct TLP.
- Operate the unit's crew-served weapons.
- Conduct urban reconnaissance and combat patrolling.
- Enter and clear buildings and rooms as part of an urban attack or cordon and search operation, which includes—
  - Sensitive site exploitation (SE).
  - Use of metal detectors.
  - Employment of military working dogs.
  - Tactical call out.
  - Collaboration with local army, police, or special operations forces.
- Defend an urban area.

- Act as a member of a mounted patrol including specific driver training.
- Recover own vehicles.
- Control civil disturbances.
- Navigate in an urban area.
- Prepare for follow-on missions.
- Identify explosives, bombs, booby traps, their construction materials, and methods for making and clearing them.
- Link up with the battlespace owner.
- React to contact, ambush, snipers, indirect fire, and improvised explosive devices (IEDs).
- Set up a personnel or vehicle checkpoint or blocking positions around a target location.
- Establish overwatch positions and support by fire positions such as sniper positions.
- Simultaneously clear the top and bottom floors of a building.
- Assign climbing and roof-clearing teams for overwatch or sniper support.
- Teach the effective use of long-range surveillance, scout, and sniper teams.
- Secure a disabled vehicle or downed aircraft.
- Call for indirect fire and CAS.
- Create and employ explosive charges.
- Handle detainees and EPWs and extract high-value targets.
- Treat and evacuate casualties.
- Accurately report information.
- Understand the society and cultures specific to the AO.
- Use basic commands and phrases in the region's primary language.
- Conduct tactical questioning (TQ).
- Interact with media personnel.
- Conduct thorough after action reviews.

ANALYZING THE URBAN ENVIRONMENT

11-8. Urban operations often differ from one operation to the next. However, some fundamentals apply to urban operations regardless of the mission, geographical location, or level of command. They are particularly relevant to the urban environment, which is dominated by artificial structures and a dense noncombatant population. These fundamentals help ensure every action taken by a commander conducting urban operations contributes to the desired end state.

11-9. Maintaining close combat is inherent in decisive action urban operations. Close combat in any urban operation is resource intensive, requires properly trained and equipped forces, and has the potential for a high number of casualties. The ability to close decisively with and destroy enemy forces as a combined arms team remains essential. In stability urban operations, lack of respect for and fear of Army forces can hinder recovery as much as the ill-advised use of force. Ideally, all brigade combat team Soldiers are properly equipped and trained to fight in an urban environment. This readiness allows the brigade combat team to deter aggression, compel compliance, morally and physically dominate an enemy and destroy their means to resist, and terminate or transition urban operations on the brigade combat team commander's terms.

11-10. Previous Army doctrine inclined toward a systematic, linear approach to urban combat. This attrition approach emphasized standoff weapons and firepower. It can result in significant collateral damage, a lengthy operation, and inconsistency between the political situation and strategic objectives. Enemy forces defending urban areas want Army forces to adopt this approach due to the likely costs in resources. Brigade combat team commanders consider this approach to urban combat only as an exception justified by unique circumstances. Instead, commanders seek to achieve precise, intended effects against multiple decisive points that overwhelm an enemy's ability to react effectively.

CONTROL THE ESSENTIAL AND MINIMIZE COLLATERAL DAMAGE

11-11. Rangers need to analyze the urban environment carefully. Considerations involve—
- Mission. Know the correct task organization to accomplish the mission (offense, defense, or stability and support operations).
- Enemy:
  - Disposition — Analyze the array of enemy forces in and around the objective, those known and suspected such as known or suspected locations of minefields, obstacles, and strong points.
  - Composition and strength — Analyze the enemy's task organization, troops available, suspected strength, and amount of support from the local civilian population based on intelligence estimates. Is the enemy a conventional or unconventional force?
  - Morale — Analyze the enemy's current operational status based on friendly intelligence estimates. Is the enemy well supplied? Have they recently won against friendly forces or sustained many casualties? What is the current weather?
  - Capabilities — Determine what the enemy can employ against friendly forces. What weapon systems do they have? Are there snipers? What about threats from IEDs, high-yield explosives, and chemical, biological, radiological, and nuclear (CBRN) exposure? Are there artillery, engineer, or air defense assets? Do they have thermal imaging devices or NVDs, CAS, or armor threats? Discriminate between threats and nonthreats such as suicide vests.
  - Probable courses of action — Based on friendly intelligence estimates, determine how the enemy will likely fight in the AO (in and around the friendly AO). Know the enemy AO tactics, techniques, and procedures such as trip wires, pressure plate IEDs, or snipers. Analyze historical data from attacks (where, what, how, and time of day).

Terrain

11-12. Leaders conduct a detailed terrain analysis of each urban setting, considering the types and composition of existing structures. When analyzing terrain in and around the AO, they use: (1) observation and fields of fire, AAs, key terrain, obstacles, and cover and concealment; (2) political, military, economic, social, information, infrastructure, physical environment, and time; (3) areas, structures, capabilities, organizations, people, and events. (See chapter 2 for information on operations.) Further considerations specifically for observation and fields of fire and for cover and concealment follow:
- Observation and fields of fire — Remain ready to conduct urban operations under limited visibility conditions.
- Cover and concealment:
  - Thoroughly analyze areas in and on the edge of urban areas.
  - Identify routes to objectives that give assault forces the best possible cover and concealment.

- Take advantage of limited visibility, which allows forces to move undetected to their final assault or breaching positions.
- Use overwatch elements and secondary entry teams for security while initial entry or breaching teams move forward.
- When in the final assault position, forces move as rapidly as tactically possible to access structures, which afford cover and concealment.
- It is human nature to stick together and seek safety, but avoid bunching up at entry points, tunnels, walls, or indoors. This reduces the possibility of one grenade's impacting the entire team. Maintain a safe but securable distance between teams and squads.
- Learn to use obscurants properly and use tactical patience to take full advantage of these effects.
- Practice noise and light disciplines. Avoid unnecessary vocal communication, learn the proper use of white light, and limit contact with surfaces capable of drawing the enemy's attention.

Obstacles, Key Terrain, and Avenues of Approach

11-13. Many artificial and natural obstacles exist on the periphery, as well as in the urban environment. Conduct detailed reconnaissance of routes and objectives including subterranean complexes and consider route adjustments and special equipment needs. Ensure routes are clear (not blocked). Avoid roads that run along or through marketplaces since they can easily become blocked.

11-14. Analyze which buildings, intersections, bridges, landing zones (LZs), pickup zones (PZs), airports, and elevated areas provide a tactical advantage to either side. The leader identifies critical infrastructure in the AO that provides the enemy with a tactical advantage in the battlespace (for example, communications centers, medical facilities, governmental facilities, facilities with psychological significance).

11-15. Consider the effects of the roads, intersections, inland waterways, and subterranean constructs such as subways, sewers, and basements. Leaders classify areas as *go*, *slow go*, or *no go* based on the navigability of the approach. Always have alternate infiltration and exfiltration routes. Keep in mind a wall can be breached to create an emergency exfiltration route.

TROOPS AND TIME

11-16. Analyze each friendly force for its disposition, composition, strength, morale, capabilities, and other appropriate qualities and condition. Leaders consider the type and size of the objective to plan the effective use of available troops.

11-17. Operations in an urban environment have a slower pace and tempo. Leaders consider the amount of time required to secure, clear, or seize the urban objective, along with the stress and fatigue Rangers often experience. Leaders allow additional time for area analysis efforts. Such analysis reviews or asks the following, respectively:

- Maps, urban plans, and aerial photographs.
- Historical data collected from other units and indigenous forces.
- Hydrological data.
- Line of sight surveys.
- Long-range surveillance and scout reconnaissance.
- Is artillery supporting multiple units at the same time?
- How long does it take to shift a 155-mm howitzer and prepare the gun?

- What is the priority level for getting Armor assets?
- How close is Armor to the target?
- Does Armor's presence compromise the mission?
- How long will it take Armor to move to a location?
- If Armor assets have not been previously coordinated, how long does it take to get them?
- How much preparation, survey, and emplacement time of charges do the engineers need?

CIVILIANS

11-18. The National Command Authority establishes the ROE. Commanders at all levels may provide further guidance for dealing with civilians in the AO. Leaders remind subordinates daily of the latest ROE and immediately inform them of any changes. Rangers have the discipline to identify the enemy from noncombatants and ensure civilians understand and follow all directed commands.

Note. It is possible civilians do not speak English, are hiding (especially small children), or are dazed from a breach. Do not give them the means to resist. Rehearse how clearing or search teams react to these variables. Never compromise the safety of the Rangers. Consider having the interpreter use a marking system to separate military-aged males from women and children. Have designated dirty and clean rooms and a tactical questioning area.

11-19. The complexity of the urban environment, particularly the human dimension, requires rapid information sharing at all levels including joint services, multinational partners, and participating government and nongovernmental agencies. The analysis of urban information necessary for refining and deepening a commander's understanding of the urban environment and its infrastructure of systems also demands collaboration among the various information sources and consumers.

CLOSE QUARTERS COMBAT

11-20. Due to the nature of a close quarters combat encounter, engagements are very close (within 10 meters) and very fast (with targets exposed for a few seconds only). Most of these engagements are victorious for the side who hits first and neutralizes their enemy. It is more important to knock down an enemy as soon as possible than it is to kill them. To win a close quarters engagement, Rangers make quick, accurate shots by mere reflex. Their reflexive fire training allows them to accomplish this. Remember: always fire until the enemy goes down. Rangers conduct all reflexive fire training with their eyes open.

Note. Research suggests only three out of ten people actually fire their weapons when confronted by an enemy during room-clearing operations. Close quarters combat success for the Ranger begins with psychological preparedness for close quarters battle. The foundation for this preparedness begins with proficiency in basic rifle marksmanship. Survival in the urban environment does not depend on advanced skills and technologies. Rangers are proficient in the basics.

REHEARSALS

11-21. Similar to the conduct of other military operations, leaders need to designate time for rehearsals. Urban operations require a variety of individual, collective, and special tasks not associated with operations on less complex terrain. These tasks require additional rehearsal time for clearing, breaching, obstacle reduction, CASEVAC, and support teams. Additionally, leaders identify time for rehearsals with combined arms elements.

11-22. In a stance, the feet are shoulder-width apart with the toes pointed straight to the front (direction of movement). The firing side foot is slightly staggered to the rear of the nonfiring foot. The knees are slightly bent, and the upper body leans slightly forward. The shoulders are not rolled or slouched. The weapon is held with the buttstock in the pocket of the shoulder and with a firm, rearward pressure into the shoulder. This allows for more accurate shot placement on multiple targets. The firing side elbow is kept in against the body, and the hand remains forward on the weapon (not on the magazine well). This allows for better control of the weapon. The Ranger assumes a comfortable boxer stance.

11-23. For the low carry technique, the buttstock of the weapon is placed in the pocket of the shoulder. The barrel is pointed downward so the front sight post and day optic are just outside the Ranger's field of vision. The head is always up, identifying targets. This technique is safest and recommended for use by the clearing team once inside the room.

11-24. For the high carry technique, the buttstock of the weapon is held in the armpit. The barrel is pointed slightly upward with the front sight post in the peripheral vision of the Ranger. To assume the proper firing position, push out on the pistol grip, thrust the weapon forward, and pull the weapon straight back into the pocket of the shoulder. This technique is best suited for the lineup outside the door. Exercise caution with this technique, always maintaining situational awareness, particularly in a multifloored building.

Note. Muzzle awareness is critical to the successful execution of close quarters operations. Rangers never at any time point their weapons at or cross the bodies of their fellow Rangers. They always avoid exposing the muzzle of their weapons around corners, which is known as flagging.

11-25. For a weapon malfunction during any close quarters combat training, the Ranger takes a knee to conduct immediate action. Upon clearing the malfunction, the Ranger need not stand to engage targets immediately. Rangers save precious seconds by continuing to engage from one knee. Whenever other members of the team see a Ranger down, they automatically clear that Ranger's sector of fire. Before standing, the Ranger warns team members of the movement and only rises after checking the rear, verifying no one is shooting overhead, and receiving the team's acknowledgement of the warning. Whenever a malfunction occurs after a Ranger has committed to a doorway, that Ranger enters the room far enough to allow those following to enter and then moves away from the door. Rangers continue to practice this drill until it is second nature to them.

11-26. Rangers give special consideration to the approach to a building or breach point. One trademark of Ranger operations is the intentional use of limited visibility conditions. Whenever possible, Rangers execute breaching and entry operations during hours and conditions of limited visibility. Rangers always take advantage of all available cover and concealment when approaching breach and entry points. When natural or artificial cover and concealment is unavailable, Rangers employ obscurants to conceal their approach. Obscurants also enhance existing cover and concealment. Rangers number the breach and entry team members for identification, communication, and control purposes. A clearing team employs four Rangers as follows:

- Ranger #1 is always the most experienced and mature member of the team, other than the TL. Ranger #1 is responsible for frontal and entry and breach point security.
- Ranger #2 is directly behind Ranger #1 in the order of movement and moves through the breach point in the opposite direction from Ranger #1.
- Ranger #3 moves in the opposite direction as Ranger #2 inside the room, at least 1 meter (3 feet) from the door.
- Ranger #4 moves in the opposite direction as Ranger #3, is responsible for rear security, and is normally the last Ranger into the room. Ranger #4 has the additional duty of breaching.
- The TL initiates all vocal and physical commands and exercises situational awareness at all times with respect to the task, friendly force, and enemy activity. The TL maintains a position to control the team.
- With the possibility of civilians in the building or rooms, Rangers sometimes decide to enter with precision weapons only such as M4s (not M249s) to avoid civilian casualties.

Note. Consider how much firepower each Ranger delivers and where best to put the squad automatic weapon gunner in the order. Weigh firepower against quick, accurate shots. When Ranger #4 has breaching responsibilities, the squad automatic weapon gunner does not serve in that position since that reduces the firepower.

11-27. Rangers give consideration to <u>actions outside the point of entry</u>. The entry point position and individual weapon positions are important. The clearing team members stand 1 to 2 feet (about 0.5 meters) from the entry point, ready to enter. They orient their weapons to provide their own 360-degree security at all times. Team members signal to each other that they are ready at the point of entry, which is best accomplished by sending up a squeeze (rocking motion). When the team uses a tap method instead, they tend to mistake an inadvertent bump for a tap.

11-28. To <u>enter and clear a room</u>, see battle drill 07-SQD-D9509 in chapter 8. Figures 11-1 and 11-2 and figure 11-3 on page 11-10 depict how Rangers clear a room.

11-29. The procedural steps to <u>lock down a room</u> follow.
    a. Control the situation in the room.
    b. Use clear, concise hand and arm signals. Keep vocal commands to a minimum to reduce the amount of confusion and to prevent any enemy in the next room from discerning what is happening. This enhances the opportunity for surprise and allows the assault force to detect any approaching force.
    c. Physically and psychologically dominate the room's occupants.
    d. Assess the situation. Slow-clearing instead of dominating a room with brute force is sometimes more effective in a less hostile situation. It keeps noncombatants calm and more manageable.
    e. Establish security and report status.
    f. Do a cursory search of the room including the ceiling (three-dimensional fight).
    g. Identify those deceased using reflexive response techniques such as an eye thump or a kick to the groin for males.

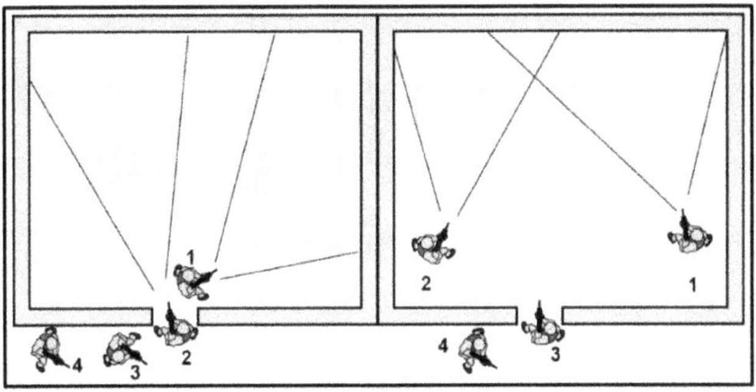

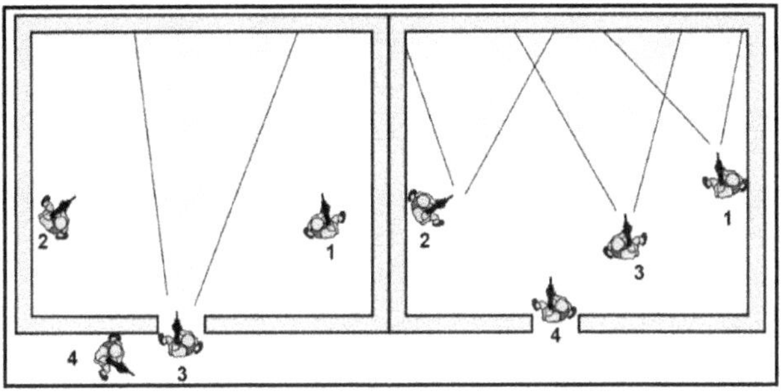

```

.

h. Search the room for the PIR while considering the time available on target.
i. Evacuate personnel.
j. Mark the room as clear in accordance with the unit's SOP (for example, with chemical light sticks, VS-17 signal panel, engineer tape).

Figure 11-1. First two Soldiers entering to clear a room

Figure 11-2. Third soldier entering room and clearing sector
```

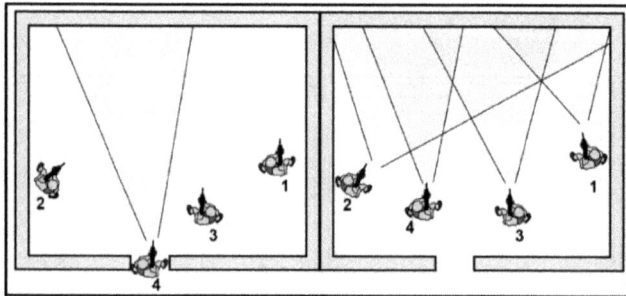

Figure 11-3. Fourth Soldier entering room and dominating sector

MARKING BUILDINGS AND ROOMS

11-30. Units have long identified a need to mark specific buildings and rooms during urban operations. Sometimes, rooms require marking for having been cleared, or buildings requiring marking for containing friendly forces. Chalk is the most common marking material; it is light, easy to obtain, and less visible than the alternatives. Other choices are spray paint and paintball guns. (See ATP 3-06.11 for information on building markings.)

Note. Avoid permanently marking buildings and rooms since this causes collateral damage, thereby potentially deteriorating relationships with local nationals.

11-31. Chemical light sticks and scrim-backed, pressure-sensitive tapes (for example, 100 mph tape) are available in a variety of colors, visible from a distance, and removable. The colors of the chemical light sticks and 100 mph tapes have different meanings:
- Red — CCP.
- Green — room clear.
- Orange — unexploded ordnance.
- Blue — clean room.
- Infrared — breach point.

Chapter 12

Waterborne Operations

This chapter discusses rope bridges, poncho rafts, and watercraft. While conducting waterborne operations, all Rangers wear the waterborne uniform. Furthermore, Rangers wear the equipment in the following order:

1. Unbloused pant leg.
2. Fastened top collar (all the way up).
3. Fastened cuffs.
4. Swimmer safety line tied with an around-the-waist bowline, an end-of-rope bowline at arm's length, and carabiners attached to the collar.
5. Unzipped field load carrier.
6. Tactical assault panel with unbuckled back buckle (stream crossing).

ROPE BRIDGE

12-1. Rope bridges are useful when a battalion-sized or smaller unit is conducting a covert gap crossing. A covert crossing is a planned crossing of an inland water obstacle or any other gap crossing during which the unit intends to remain undetected. A covert gap crossing serves a variety of situations in support of various missions but is considered (as opposed to deliberate or hasty) only when the need or opportunity arises to cross a gap without the enemy's detection. Swift river currents influence the selection of a crossing method. (See paragraph 12-36 for river current velocity.)

12-2. The Ranger patrol seldom has ready-made bridges, so the patrol team knows how to employ covert gap crossing techniques. The requisite personnel for making a rope bridge follow:

- Ranger #1 — lead safety swimmer and far side lifeguard.
- Ranger #2 — rope swimmer (swims the water obstacle pulling a 150-foot rope and ties off the rope on the far side anchor point).
- Ranger #3 — near side lifeguard and the last Ranger to cross the water obstacle.
- Ranger #4 — BTC and the team's most knowledgeable person.
- Rangers #5 and #6 — rope pullers (assist the BTC in tightening the rope bridge).

12-3. A wet crossing (or one-rope bridge) requires special equipment. This includes—

- Two carabiners for each piece of heavy equipment.
- Three steel carabiners for each 150 feet (45 meters) of rope.
- One 7-foot utility rope for each person (swimmer safety line).
- Two carabiners for each person (one clipped to the swimmer safety line and one tied to the top center frame of the rucksack).
- Two waterproof bags for each RTO.
- Two carbon dioxide inflatable life preservers (Scout swimmer vests).
- Three noninflatable life preservers (work vests).
- Two 150-foot nylon ropes.

12-4. Leaders prepare a gap crossing annex (see table 2-1 on pages 2-32 through 2-34) with the unit OPORD. They also accomplish special organization at this time. For a platoon-sized patrol, leaders normally give a squad the task of providing the rope bridge team. The SL designates the squad's most technically proficient Ranger as the BTC. Rehearsals and inspections take place. (See figure 12-1.) Leaders emphasize—

- Security and actions on enemy contact.
- Actual construction of the rope bridge on dry land within the 8-minute time standard.
- Individual preparation.
- Order of crossing.
- All signals and control measures.
- Reorganization.

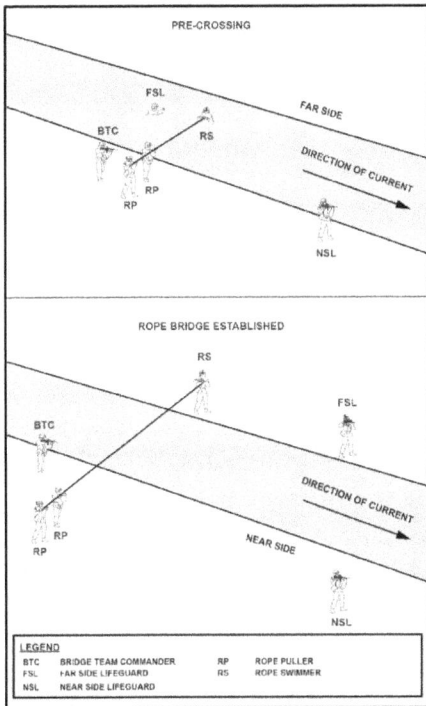

Figure 12-1. Positions of bridge team personnel

12-5. The BTC rehearses the bridge team as realistically as possible during the TLP. Leaders ensure the personnel are proficient in the mechanics of a covert gap crossing operation. They identify weak swimmers, inspect equipment, correct rigging and preparation, finalize weapon configurations, confirm personnel knowledge and understanding of the operation, and ensure the 150-foot rope is back-stacked and properly coiled. The unit leader selects the crossing site that best facilitates the tactical plan.

WARNING

The first Ranger to cross the fully constructed rope bridge tests for proper rope tension with full body weight and a 45-pound rucksack by locking elbows atop the center of the rope. The Ranger should submerge no deeper in the water than the level of the chest or nametape.

Only one special item of equipment may be on the rope bridge at a time. Special items are M240s, M249s, and additional rucksacks. The special item must completely clear the rope bridge on the far side before a Ranger may follow it. The rope bridge commander shall await the removal of the special item before attaching the next Ranger for crossing.

Any Rangers identified as weak swimmers cross alone to enable the near side and far side lifeguards to watch them without distraction.

See chapter 9 for information on rope bridges.

12-6. Once the mission reaches the execution phase, several actions take place. These include establishing and conducting a bridge for the gap crossing. The process follows.

    a. The leader halts short of the river, establishes local security, reconnoiters the area for any enemy presence, and determines the need for and suitability of a crossing site. The leader directs the BTC to construct the bridge.

    b. The BTC constructs a one-rope bridge and selects near side and, visibility permitting, far side anchor points. To anchor him- or herself to the bridge, the BTC ties a swimmer's safety line around the waist and secures it with an overhand knot, ties the free running end of the bowline into an overhand knot, and attaches a carabineer to the loop in the knot. The bowline should be just long enough to place the carabineer at arm's length, which allows the BTC to remain within reach of the rope bridge were the BTC to lose grip.

    c. The bridge team begins to establish the rope bridge while unit members begin individual preparation.

    d. Each Ranger puts a carabineer in the end of the bowline and in the front sight assembly of every M4. M240 and M249 gunners put a carabineer through the front sling attachment point and rear sling attachment point of their machine guns. The RTO, FO, and others with heavy rucksacks place an additional carabineer on the top center of their rucksack frames.

    e. Rucksacks are waterproofed to protect the equipment inside and to ensure the rucksack remains buoyant in the water and can serve as a flotation device. Rucksack straps remain tight to prevent the rucksack from pushing the Soldier's head forward into the water when negotiating the water obstacle.

    f. The team establishes security upstream and downstream while the unit leader briefs the BTC on anchor points. The BTC counts the Rangers across.

    g. The BTC enforces noise and light disciplines and maintains security.

    h. The order of Soldiers crossing is based on the METT-TC (I) variables and the locations of key weapons and leaders in case the platoon makes contact while separated by the water obstacle.

12-7. The bridge team is responsible for constructing the rope bridge. Ranger #1 (lead safety swimmer and far side lifeguard) grounds the rucksack (with a carabineer through the top of the frame) to the rear of the near side anchor point and carries a knotted hand line or safety line to assume duties of the far side lifeguard. Ranger #2 (rope swimmer) grounds the rucksack in the same manner as Ranger #1.

Note. Wear equipment in the following order (from the body out):

Waterborne uniform (fastened top and cuffs, unbloused pants).
Noninflatable life preserver.
Carbon dioxide inflatable life preserver.
Field load carrier / Tactical assault panel.
Weapon (across the back).
Swimmer safety line routed over all equipment and secured to the collar of the Army uniform blouse.

12-8. Ranger #1 enters the water upstream from Ranger #2 and stays an arm's length from Ranger #2. Ranger #1 identifies the far side anchor point upon exiting the water. Ranger #1 disconnects the rope from Ranger #2 upon exiting the water and then moves to the far side lifeguard position downstream of the rope bridge with knotted hand line (in nonthrowing hand), field load carrier / tactical assault panel, weapon grounded, and noninflatable life preserver (in throwing hand). Ranger #1 continues to wear the carbon dioxide inflatable life preserver.

12-9. Ranger #2, in the same type of waterborne uniform as Ranger #1, has the duties to swim across the water obstacle while pulling the rope and to tie off the rope on the far side anchor point identified by Ranger #1. Initially, Ranger #2 wraps the rope one time around the far side anchor point and then clips the snap link back upon the rope. Ranger #2 signals the BTC when this is complete. Once the BTC has tied the figure eight slip on a bight of the transport-tightening system on the near side anchor point, Ranger #2 pulls all the slack to the far side and secures the rope by wrapping the rope around the anchor 18 to 24 inches (46 to 61 centimeters) above the water with a minimum of four turns (the number of turns being situationally dependent) and then attaching the snap link back upon the rope. (See figure 12-2 on page 12-6.) The direction of the round turns is the same direction as the flow of the water current to facilitate exiting off the rope bridge.

Note. Wear equipment in the following order (from the body out):

Waterborne uniform (fastened top and cuffs; unbloused pants).
Noninflatable life preserver.
Field load carrier / Tactical assault panel.
Weapon (across the back).
Swimmer safety line tied with an around-the-waist bowline. The carabiner of the swimmer safety line routed through an end-of-line bowline no more than an arm's length and then secured by reattachment to the swimmer safety line routed around the rope swimmer's waist (in the vicinity of the small of the back).

Figure 12-2. Example of tensionless, natural anchor for far side anchor point

12-10. Ranger #3 (near side lifeguard) takes position on the downstream side of the near side anchor point before Rangers #1 and #2 enter the water. Ranger #3 wears the same type of waterborne uniform as Ranger #1. Ranger #3 grounds the rucksack (with a carabineer through the top of the frame) on the rear of the near side anchor point. After the PSG crosses and verifies the headcount, Ranger #3 unties the quick release at the near side anchor point. Ranger #3 is the last Ranger pulled across the water obstacle. Before crossing the water obstacle, Ranger #3 dons equipment in the following order:

- Waterborne uniform (fastened top and cuffs; unbloused pants).
- Noninflatable life preserver.
- Carbon dioxide inflatable life preserver.
- Field load carrier / Tactical assault panel.
- Weapon (across the back).
- Swimmer safety line routed over all equipment and secured to an end-of-line bowline at the end of the rope.

12-11. Ranger #4 (BTC) wears the standard waterborne uniform with the field load carrier / tactical assault panel and a sling rope as a safety line tied in an around-the-waist bowline with an end-of-line bowline no more than an arm's length. Ranger #4 is responsible for the organization of the bridge team, construction of the rope bridge, back-feeding the rope, and tying the transport-tightening system (see figure 12-3 on page 12-8). The BTC designates the near side anchor point and, upon receiving the signal from Ranger #2 that the far side anchor point is secure, ties the figure eight slip on a bight of the transport-tightening system 18 to 24 inches (46 to 61 centimeters) above the water, performing one single turn around the near side anchor point and clipping into the two interlocking carabineers. The BTC directs Ranger #2 to remove all slack and secure the rope on the far side anchor point. Once Ranger #2 signals again that the far side anchor point is secure, the BTC directs Rangers #5 and #6 to pull and tighten the rope. Once the rope is tight, the BTC wraps it around the near side anchor point with a minimum of four turns (the number of turns being situationally dependent) and secures it with a half hitch and a quick release. The BTC ensures the transport-tightening knot is on the upstream side of the rope bridge, all individuals are in the waterborne uniform, all equipment is properly configured, rucksack straps are cinched tight (which prevents a rucksack from riding forward and pushing the Soldier's head into the water when negotiating the obstacle), and personnel are hooked into the rope and facing upstream. The BTC ensures all weapons are attached to the rope and controls the flow of personnel on the rope bridge. The BTC is responsible for crossing with the rucksack of Ranger #1 and is generally the next-to-last Ranger to cross, following the PSG, who is keeping a headcount.

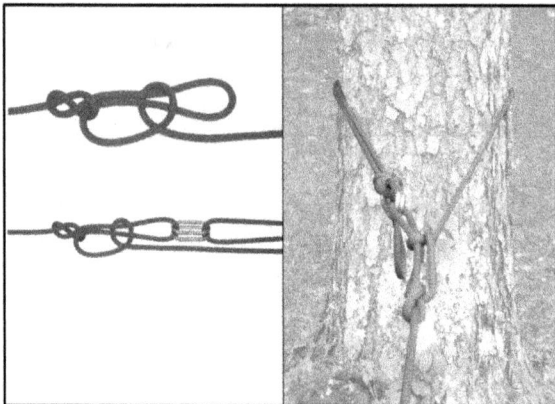

Figure 12-3. Example of transport-tightening system for near side anchor point

12-12. Rangers #5 and #6 (rope pullers) wear the waterborne uniform with the field load carrier / tactical assault panel and safety line. They tighten the transport-tightening system. They also take across the rucksacks of Rangers #2 and #3. Once they reach the far side, Rangers #5 and #6 pull the last Ranger (#3) across.

12-13. Rangers #4, #5, and #6 transport the rucksacks of Rangers #1, #2, and #3. To do so, they hook the rucksacks into the rope by running the carabineer through the top of the frames and then pulling the rucksacks across the bridge. They attach their own weapons between themselves and the rucksacks they are pulling.

12-14. The M240 and M249 are secured to the bridge by carabiners on the front sight post and rear swivel. The M240 is pulled across by the trailing arm of the M240 gunner.

12-15. AN/PRC-119F is waterproofed before crossing a one-rope bridge. Upon the successful establishment of far side FM communications, the near side RTO breaks down and waterproofs the radio and prepares to cross the bridge. The RTO places a carabineer in the top center of the rucksack frame (same as for Rangers #1, #2, and #3). The BTC hooks the rucksack to the rope.

Note. The use of two carabiners binds the load on the rope. Adjust arm straps all the way out. The RTO pulls the radio across the rope bridge.

12-16. Special consideration goes to <u>heavy equipment</u> when the team is crossing a water obstacle with it. Rangers pulling across heavy equipment as they themselves cross increases the hazard for those individuals. Pulling across these items with a haul line sometimes requires additional ropes and snap links.

PONCHO RAFT

12-17. Normally, a team constructs a poncho raft to cross rivers and streams when the current is not swift. A poncho raft is especially useful when the unit is still dry and when the PL wants to keep equipment dry. There are several things to consider when constructing and using a poncho raft:

- Equipment requirements:
  - Two serviceable ponchos.
  - Two weapons (or poles).
  - Two rucksacks for each team.
  - 10 feet (3 meters) of utility cord for each team.
  - One sling rope for each team.
- Conditions — Poncho rafts are useful for crossing water obstacles with at least one of these conditions: the water obstacle is too wide for a 150-foot rope; or no sufficient near or far side anchor points are available to allow for rope bridge construction. Yet, under no circumstances are poncho rafts to be used to cross a water obstacle when the current is unusually swift.
- Crossing site — Before using a crossing site, leaders conduct thorough reconnaissance of the immediate area. Analyzing the situation by using the METT-TC (I) variables, the patrol leader chooses a crossing site that offers as much cover and concealment as possible and has entrance and exit points as shallow as possible. For speed of movement, the best choice is a crossing site with easily traversed near and far side banks.
- Execution phase — The procedure to construct a poncho raft assumes Rangers have undamaged ponchos that will remain watertight. This procedure follows:
  - Pair off the unit or patrol to have the necessary equipment.
  - Tie off the hood of one poncho and lay it out on the ground with the hood up.
  - Place weapons, muzzle to butt, in the center of the poncho about 18 inches (45 centimeters) apart.
  - Place rucksacks and field load carriers / tactical assault panels between the weapons with the rucksacks as far apart as possible.
  - Start to undress, bottom to top and boots first (weather permitting). When necessary to serve as the tie downs in subsequent steps, pull laces out of boots.
  - Place boots over the muzzle or butt of a weapon with the toe in.
  - Continue to undress, folding each item neatly and placing clothes items on top of boots.
  - Once all the equipment is between the two weapons or poles, snap the poncho together.
  - Lift the snapped portion of the poncho into the air and tightly roll it down to the equipment. Start at the center and work out to the end of the raft, creating pigtails at the end. Two Rangers working together accomplish this more quickly and easily.
  - Fold the pigtailed ends inward and tie them off (with a single bootlace as necessary).
  - Lay out the other poncho on the ground with the hood up.

- In the center of this poncho, place the other poncho with equipment.
- Snap, roll, and tie the whole package as before. Tie utility cord (or the third and fourth bootlaces as necessary) around the raft about 1 foot (30 centimeters) from each end for added security. The poncho raft is now complete.

Note. The patrol leader analyzes the situation by using the METT-TC (I) variables and decides the appropriate uniform for crossing the water obstacle, such as placing weapons inside the poncho raft versus slinging them across the back or remaining dressed versus stripping down and placing clothes inside the raft.

WATERCRAFT

12-18. The use of inland and coastal waterways often adds flexibility, surprise, and speed to tactical operations. The use of these waterways also increases the load-carrying capacity of normal dismounted units. Watercraft are employed in reconnaissance and assault operations.

12-19. A waterborne insertion annex (see table 2-14 on pages 2-30 through 2-32) undergoes preparation with the unit OPORD, and the accomplishment of special organization occurs at this time. The PL designates the most technically proficient Rangers as coxswains.

12-20. The combat rubber raiding craft is a lightweight, ten-person, inflatable watercraft, useful on inland and coastal waterways. There are five pressure relief valves inside the buoyancy tubes and five separate airtight compartments. To pump air into the boat, turn all the valves to the INFLATE (orange) section of each valve. Once the assault boat is full of air, turn all the valves to their NAVIGATION (green) sections. This partitions the assault boat into eight separate compartments. The characteristics of the watercraft follow:

- Maximum carrying capacity — 2,760 pounds (1,252 kilograms) including engine, personnel, equipment, fuel, and deck.
- Crew — 10 (not to exceed maximum carrying capacity).
- Overall length — 15 feet, 5 inches (4.7 meters).
- Overall width — 6 feet, 6 inches (2 meters).
- Weight with an aluminum rollup floor — 304 pounds (138 kilograms).
- Weight with a composite rollup — 274 pounds (124 kilograms).
- Weight with a hard deck — 285 pounds (129 kilograms).
- Weight without a deck — 210 pounds (95 kilograms).
- Recommended 40 horsepower / maximum 65 horsepower (maximum engine weight 280 pounds [127 kilograms]).

PREPARATION, PERSONNEL, AND EQUIPMENT

12-21. Units lash crew-served weapons, radios, ammunition, and other bulky equipment securely to the boat to prevent loss were the boat to overturn. They ensure machine guns that had hot barrels are cool prior to lashing them inside the boat. See figure 12-4 for rucksack tie down and figure 12-5 on page 12-12 for rigging equipment in the boat. These are specific preparations, personnel, equipment, and procedures associated with watercraft:

- The rubber boat:
  - Each rubber boat has a 12-foot bowline secured to the front starboard D-ring. Tie this rope with an anchor line bowline and cover the knot with 100 mph tape.
  - Each rubber boat has a 15-foot centerline secured to the rear floor D-ring. Use the same procedure for securing the centerline as for the bowline.
  - Each rubber boat has 4 pounds per square inch (276 millibars) of air. Check that all valve caps are tight and set in the NAVIGATE position.
  - Each rubber boat has one foot pump, which is placed in the boat's front pouch or, when no pouches are present, on the floor.
  - Use the maintenance chart to inspect each rubber boat as appropriate.
- Personnel and equipment:
  - All personnel wear work vests, kapok, or another suitable positive flotation device.
  - Personnel wear the field load carrier / tactical assault panel over the work vest and unbuckled at the waist.
  - Personnel sling their individual weapons across the back with muzzle pointed downward and facing toward the inside of the boat.
  - Waterproof radios and batteries.
  - Pad pointed objects to prevent their puncturing the boat.

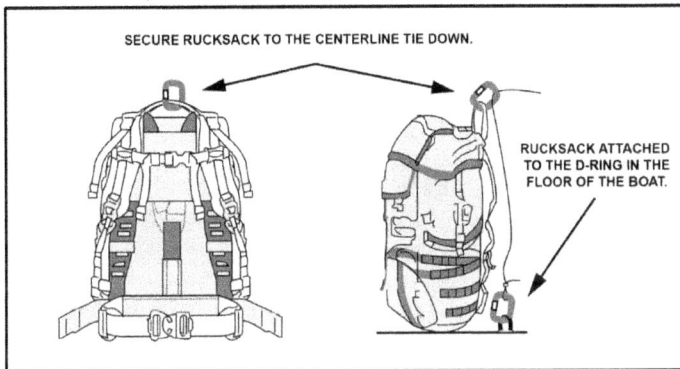

SECURE RUCKSACK TO THE CENTERLINE TIE DOWN.

RUCKSACK ATTACHED TO THE D-RING IN THE FLOOR OF THE BOAT.

Figure 12-4. Rucksack rigging

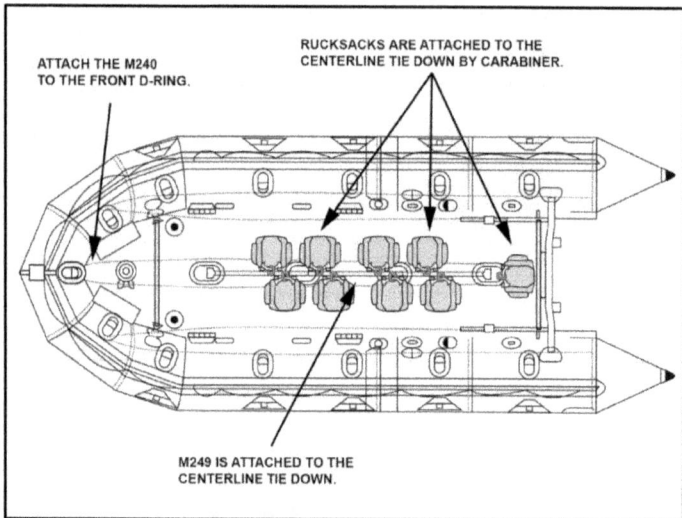

Figure 12-5. Equipment rigging

12-22. When rigging weapons to be lashed to the boat, attach two carabineers to the front and rear of the M240 with a middle-of-the-line bowline connecting the two carabineers. Connect a third carabineer to the center of the rope. The M249 machine gun also has two carabineers attached to the front and rear with parachute cord attached to both carabineers. (See figure 12-6.) Rucksacks have one carabineer attached to the top of each pack. (See figure 12-4 on page 12-11.)

M240 RIGGING

TWO ENDS OF BOWLINE ATTACHED TO FRONT AND REAR OF WEAPON WITH CARABINER.

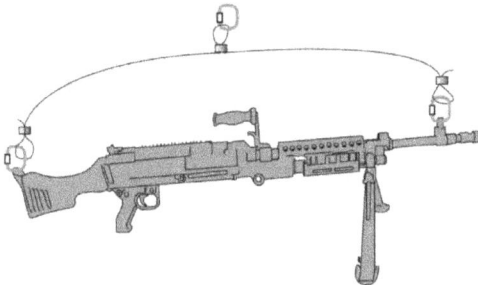

M249 RIGGING

CARABINERS ATTACHED TO FRONT AND REAR TO SECURE WEAPON ONTO CENTER OF TIE DOWN.

Figure 12-6. Weapon rigging

12-23. Place the rucksacks in the boat with the frames facing inward and tie them down through the carabineers attached to the tops of the packs. Tie the end of the centerline near the bow to the left or right D-rings on the buoyancy tube with a round turn and two quick releases. (See figure 12-5 on page 12-12.) Attach the M240 machine gun to the D-ring at the bow and the M249 machine gun to the centerline by the carabineers, ensuring the weapon is on top of the rucksacks.

12-24. Each Ranger has a specific, assigned boat position (see figure 12-7), and all have various duties as well as embarkation and debarkation procedures.

    a. Duties.

        (1) Designate a commander, who also normally serves as coxswain, for each boat.

        (2) Designate a navigator, normally a leader within the platoon, and any necessary observer team.

        (3) Position the crew as figure 12-7 depicts.

        (4) Duties of the coxswain.

            (a) Control the boat and actions of the crew.

            (b) Supervise the loading, lashing, and distribution of equipment.

            (c) Maintain the course and speed of the boat.

            (d) Give all commands.

        (5) Paddler #2 (long count) is responsible for setting the pace.

        (6) Paddler #1 is the observer, stowing and using the bowline unless another observer is assigned.

    b. Embarkation and debarkation procedures

        (1) When launching, the crew maintains a firm grip on the boat until they are inside it. When beaching or debarking, the crew holds onto the boat until it is completely out of the water. They load and unload using the bow as the entrance and exit point.

        (2) The crew keeps a low center of mass when entering and exiting the boat to avoid capsizing it. They maintain three points of contact at all times.

        (3) The long count is a method of loading and unloading by which the boat crew embarks or debarks individually over the bow of the boat. It is useful at riverbanks, on loading ramps, and when deep water prohibits the use of the short count.

        (4) The short count is a method of loading or unloading by which the boat crew embarks or debarks in pairs over the sides of the boat while the boat is in the water. It is useful in shallow water and allows the boat's quick extraction from the water. Surf operations are the primary application of this method of organization.

        (5) Beaching the boat is a method of debarking the entire crew at once into shallow water and their quick extraction of the boat out of the water.

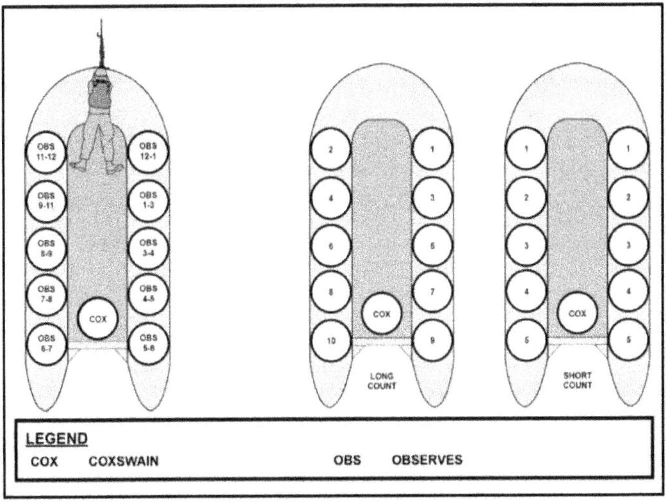

Figure 12-7. Crew positions – long count and short count

12-25. The coxswain issues <u>commands</u> for the transportation of the boat over land and its control in the water. All crewmembers learn and react immediately to all the coxswain's commands. The various commands follow:

- SHORT COUNT. COUNT OFF. — The crew counts off their positions by pairs such as one, two, three, four, five, coxswain.
- LONG COUNT. COUNT OFF. — The crew counts off their positions by individual such as one, two, three, four, five, six, seven, eight, nine, ten, coxswain.
- BOAT STATIONS. — The crew takes position alongside the boat.
- HIGH CARRY. MOVE. — Useful for long-distance moves over land, whose conduct follows:
  - On the preparatory command of HIGH CARRY, the crew faces the rear of the boat and squats down, grasping the carrying handles with their inboard hands.
  - On the command MOVE, the crew swivels, lifting the boat to their shoulders so the crew is standing and facing toward the front with the boat on their inboard shoulders.
  - The coxswain guides the crew during movement.

- LOW CARRY. MOVE. — Useful for short-distance moves over land, whose conduct follows:
  - On the preparatory command of LOW CARRY, the crew faces the front of the boat, bends at the waist, and grasps the carrying handles with their inboard hands.
  - On the command of MOVE, the crew stands erect and raises the boat about 6 to 8 inches (15 to 20 centimeters) off the ground.
  - The coxswain guides the crew during movement.
- LOWER THE BOAT. MOVE. — The crew lowers the boat gently to the ground by the carrying handles.
- GIVE WAY TOGETHER. — The crew paddles to the front with paddler #2 setting the pace.
- HOLD. — The entire crew keeps their paddles straight downward and motionless in the water to stop the boat.
- LEFT SIDE, HOLD. — The portside crew holds, and the starboard-side crew continues with the previous command.
- BACKPADDLE. — The entire crew paddles backward, propelling the boat to the rear.
- BACKPADDLE, LEFT. — The portside crew backpaddles causing the boat to turn left, and the starboard-side crew continues with the previous command.
- REST PADDLES. — Crewmembers place their paddles on their laps with the blades outboard. The coxswain may give this command in pairs (for example, #1s, REST PADDLES).

CONDUCT CAPSIZE PROCEDURES

> **WARNING**
>
> During training in cold weather months, the unit should adhere to the water submersion chart. The unit should provide a safety boat for each boat team conducting this task. All personnel shall wear the waterborne uniform and a serviceable noninflatable life vest.
>
> Rig boats for potential capsizing. The coxswain ensures the securement of all equipment. This sometimes requires additional tie downs.

12-26. The capsize drill prepares Rangers to safeguard lives and equipment were the boat to overturn.

TASK: Conduct Capsize Procedures

CONDITIONS: The platoon is conducting a waterborne insertion using the combat rubber raiding raft. The platoon has organized into ten-person boat teams. The platoon has all the requisite safety equipment; all platoon-organic equipment is tied down in accordance with the unit's SOP. The unit has communications with higher and adjacent units. Conduct some iterations of this task during hours of limited visibility.

CUE: The boat team needs to capsize the boat intentionally, which becomes necessary when the boat fills with water due to rough seas or heavy rainfall.

STANDARDS: Each boat team properly rigs and lashes its boat. Each boat team intentionally capsizes its boat and then rights its boat. Each team recovers all its personnel and equipment back into the boat and continues its mission.

TASK STEPS

1. The coxswain gives the command LONG COUNT. COUNT OFF. After the long count, the coxswain gives the command PASS PADDLES. The crew passes all paddles to the rear of the boat by crewmembers raising their paddles over their heads (except paddlers #7 and #8) so the crewmembers behind them can take them. The last two crewmembers hold the paddles until the boat is righted.

2. The coxswain designates three crewmembers (#2, #4, and #6) to remain in the boat. (They capsize the boat after the others are in the water.) The coxswain then orders the other members out of the boat by commanding 1s OUT, 3s OUT, 5s OUT, until only three Rangers remain in the boat. Once out of the boat, the Rangers move about 3 meters (10 feet) away from the boat.

3. The coxswain designates Ranger #1 (in the water) to hold onto the boat so as to be pulled onto the boat once the crew capsizes it. Ranger #1 does so by holding onto two carrying handles. The three crewmembers still in the boat each grasp a capsize line attached to three D-rings. They stand and lean backward until they capsize the boat. This pulls Ranger #1 onto the boat as it capsizes.

4. The coxswain designates a Ranger to pull the quick release that attaches the centerline tie down to the D-ring on the bow of the boat. (Omit this step whenever the three crewmembers can right the boat without disconnecting the centerline quick release.)

5. Ranger #1 assists Rangers #2 and #4 onto the boat to help in righting it. Ranger #6 (in the water) holds onto the boat in order so as to be pulled onto the boat once the crew rights it. Ranger #6 does so by holding onto two carrying handles. Rangers #1, #2, and #4 each grasp a capsize line, stand, and lean backward until they right the boat. This pulls Ranger #6 onto the boat as it rights.

6. Once the boat is righted, all crewmembers move to and hold onto the boat. Ranger #6 assists other crewmembers back into the boat. Rangers #7 and #8 pass the paddles to the other crewmembers and receive assistance onto the boat.

TASK: Conduct Capsize Procedures (continued)

7. When the crew has released the quick release on the centerline tie down, crewmembers then recover the attached equipment and resecure the centerline tie down.

8. Once all the equipment and crewmembers are in position, the coxswain has everyone count off using the long count. The Rangers also check and account for their equipment. The coxswain then gives the crew the appropriate orders and continues the mission.

RIVER MOVEMENT, NAVIGATION, AND FORMATIONS

12-27. It is very important that Rangers understand the characteristics of the river and how to navigate the water using various formations. Before embarking, it is vital they know the local conditions of the river and its movement. Terminology and knowledge common to water navigation follow, respectively:

- *Bend* — A turn in the river course.
- *Reach* — A straight portion of river between two curves.
- *Slough* (pronounced SLOO) — A dead-end branch from a river, normally quite deep and distinguishable from the river by its lack of current.
- *Dead water* — A part of the river that has no current due to erosion and changes in the river's course; characterized by an excess of snags and debris.
- *Island* — Usually a pear-shaped mass of land in the river's main current whose upstream portions tend to catch debris and, therefore, avoidance.
- The current in a narrow part of a reach is normally greater than in the wide portion.
- The current is greatest on the outside of a curve; sandbars and shallow water tend to exist on the inside of the curve.
- Sandbars are located at those points where a tributary feeds into the main body of a river or stream.

12-28. Because Rangers #1 and #2 are sitting on the front port and starboard sides of the boat, they observe for obstacles as the boat moves downriver. Upon noticing any obstacle on either side of the boat, these observers notify the coxswain. The coxswain adjusts steering to avoid the obstacle.

12-29. The patrol leader is responsible for navigation. Rangers have three acceptable methods of river navigation:

- Checkpoint and general route are two. These two methods are—
  - Useful when a well-defined checkpoint marks the drop zone and the waterway is not confusing from a lot of branches and tributaries.
  - Best for daylight hours and short distances.
- Navigator observer is the most accurate means of river navigation and is effective in all light conditions. The requisite navigation equipment includes—
  - Compass.
  - Global Positioning System.

- Photo map (first choice).
- Topographical map (second choice).
- Poncho (for night use).
- Pencil or grease pencil.
- Flashlight (for night use).

12-30. The navigator sits in the center of the boat and does not paddle. During hours of darkness, the navigator uses the flashlight under the poncho to check the map. The observer (Ranger #1) is at the front of the boat. Working together—

- The navigator keeps the map and compass oriented at all times.
- The navigator keeps the observer informed about the configuration of the river by announcing bends, sloughs, reaches, and stream junctions as the map displays them.
- The observer compares this information with the actual, observable bends, sloughs, reaches, and stream junctions. As the observer confirms the features, the navigator confirms the boat's location on the map.
- The navigator also keeps the observer informed about the general azimuths of reaches the map depicts, and the observer confirms them with actual compass readings of the river.
- The navigator announces one configuration at a time to the observer and does not announce another until confirmation and completion of each.

12-31. A strip map drawn on clear acetate and backed by luminous tape has proven useful. The drawing is to scale or a schematic. It illustrates all the curves, azimuths, and distances of all reaches. It can also include terrain features, stream junctions, and sloughs.

12-32. Various boat formations are useful (day or night) for control, speed, and security. The choice of formation depends on the tactical situation and the patrol leader's discretion. Hand and arm signals are the most useful for controlling the assault boats. The formations are wedge, line, file, echelon, and vee.

SECURE THE LANDING SITE

12-33. When the patrol is approaching an unsecured landing site, a security boat lands, reconnoiters the site, and then signals the remaining boats to land as appropriate. This is the best way. When the landing site is not secured before the waterborne force landing, the patrol considers some form of early warning such as scout swimmers. In this scenario, these Rangers swim to shore from the assault boats and signal the boats to land. The patrol rehearses all signals and actions prior to the actual operation.

12-34. The assault boats secure the landing site by force by landing simultaneously in a line formation. While this is the least desirable method of securing a landing site, the patrol rehearses it in the event the tactical situation requires its use. Arrival at the debarkation point involves several steps:

- Unit members disembark in accordance with the leader's order.
- Local security is established.
- Leaders account for personnel and equipment.
- Unit continues movement as follows:
  - The Rangers pull security, initially wearing their work vests.
  - The coxswain and two Rangers unlash and derig the rucksacks.
  - The Rangers return in the buddy system or teams, secure the rucksacks, and leave the work vests.
  - The Rangers camouflage and cache the boats prior to movement as necessary.

QUARTERING PARTY PROCEDURES

12-35. A quartering party is a patrol who departs ahead of a main body. Their purpose is to secure, reconnoiter, and organize an area for the main body's arrival and occupation. During waterborne operations, the quartering party leaves early to inspect and prepare small boats such as the combat rubber raiding raft for rigging and lashing. This saves time and facilitates an expedient and tactical occupation and departure from the beach landing site. Procedures include but are not limited to the following:

- The quartering party departs ahead of the main body. The party comprises a senior leader, RTO, security element, and all coxswains.
- The quartering party issues contingency plans.
- The quartering party is counted out but maintains communication with the main body. Perimeter security undergoes readjustment.
- The quartering party arrives at the beach landing site and establishes local security.
- The party's senior leader conducts a partisan linkup to coordinate for small boats.
- Upon identification of boats, the coxswains inspect the boats for serviceability and equipment.
- The coxswains use proper commands and lifts to move the boats into position at the actual launch point.
- The coxswains ready the equipment (for example, paddles, work vests, centerline rope) for the main body's arrival.
- The main body arrives and conducts linkup with the quartering party, and security undergoes readjustment. Information disseminates among leaders.
- The coxswains begin supervising and directing the boat crews to line up rucksacks and to secure work vests.

> **WARNING**
> Coxswains must ensure boats remain afloat during loading. Equipment and personnel can cause the boat to rest on the river bottom in shallow water, which can damage the boat.

- With the assistance of a crewmember each, the coxswains begin rigging and lashing rucksacks and other heavy equipment.
- Once the crews have secured all the rucksacks and heavy equipment, the coxswains begin directing the loading of boat crews.
- Leaders pull the remaining security elements from the perimeter, but security continues while aboard to ensure no security gaps.
- The PL and PSG maintain accountability.
- The platoon is postured for boat movement.

MEASURING CURRENT VELOCITY

12-36. Determining the current's velocity is critical to effective and safe covert gap crossing. While using watercrafts, the current velocity assists leaders in determining the speed traveled for navigation and for hasty planning purposes. A reasonable estimation involves measuring a distance along the riverbank and noting the time a floating object takes to travel that distance. Dividing the distance by the time provides the current's velocity in meters per second. (See figure 12-8.)

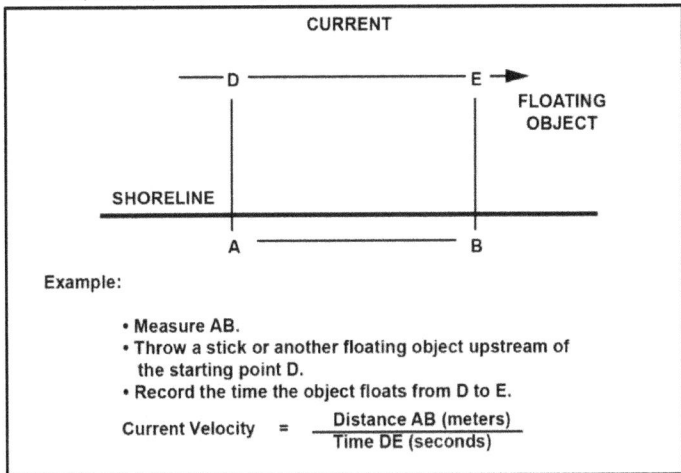

CURRENT

FLOATING OBJECT

SHORELINE

Example:

- **Measure AB.**
- **Throw a stick or another floating object upstream of the starting point D.**
- **Record the time the object floats from D to E.**

$$\text{Current Velocity} = \frac{\text{Distance AB (meters)}}{\text{Time DE (seconds)}}$$

Figure 12-8. Measuring current velocity

12-37. Correlating the desired maximum current velocity with a familiar comparative unit of measure helps in estimating the current's velocity. Below are velocities with corresponding speeds in miles per hour and in knots:

- 0.45 meters/second = 1 mile/hour = 0.87 knots.
- 0.89 meters/second = 2 miles/hour = 1.74 knots.
- 1.34 meters/second = 3 miles/hour = 2.61 knots.
- 1.79 meters/second = 4 miles/hour = 3.48 knots.
- 2.24 meters/second = 5 miles/hour = 4.34 knots.

This page intentionally left blank.

Chapter 13

Mounted Patrol Operations

This chapter outlines a technique for conducting vehicle-mounted patrol operations. Mounted patrol operations present a challenge to the Ranger leader. Trucks and other combat vehicles produce a large signature in the battlespace and increase the unit's value as a target. Vehicle movement is restricted to traversable roads and terrain. (See ATP 4-01.45 and ATP 3-21.8 for information on mounted patrol operations.)

PLANNING

13-1. When a mounted patrol will be part of the operation, leaders incorporate planning considerations for that mounted patrol while they use the eight steps of the TLP. A mission analysis using the METT-TC (I) variables includes the following information, exempting the informational considerations variable due to its excessive complexity for squad- and platoon-level operations:

- Mission — The PL extracts the following information from the company OPORD:
  - Vehicle support (number and type of vehicles and allowable combat load).
  - Weather including road conditions.
  - Vehicle pickup and dropoff location(s) and markings.
  - Vehicle movement timeline (pickup time, movement time, and other information).
  - Vehicle routes including primary and alternate checkpoints.
- Enemy:
  - Known or suspected locations in the AO or along planned routes.
  - Potential locations for ambush or IED emplacement.
  - Recent activities or reactions to mounted patrol operations.
- Terrain:
  - Potential pickup and dropoff locations.
  - Route and pickup and dropoff location evaluations using observation and fields of fire, AAs, key terrain, obstacles, and cover and concealment.
  - Weather effects and road impacts.
- Troops:
  - Number of passengers for each vehicle.
  - Chalks and chalk leaders identified.
  - Tactical cross-load.
  - Linkup and marking teams identified.
  - Pickup and dropoff security plan.
- Time — Leaders use the reverse planning sequence, which the following list represents:
  - Ground tactical plan.
  - Unloading plan.
  - Ground movement plan.
  - Loading plan.
  - Staging plan.

- Civilians:
  - ROE actions with civilians during movement.
  - ROE actions with civilian vehicles during movement.

Note. Allocate time for movement, reconnaissance, and establishing security.

13-2. There are five phases of a mounted patrol. Each phase supports the ground tactical plan, which specifies actions in the objective area to accomplish the commander's intent for the assigned mission whether a raid, ambush, reconnaissance, or other follow-on mission.

13-3. The five phases are the staging plan, loading plan, ground movement plan, unloading plan, and ground tactical plan. These involve the following tasks and conditions:

- Staging plan:
  - Establish security.
  - Employ markings and recognition signals for day and night.
  - Link up.
  - Conduct final friendly unit coordination.
  - Disseminate information and any changes to subordinate leaders.
- Loading plan for task organization and tactical cross-loading. Leaders assign each Ranger to a vehicle and ensure the tactical cross-loading of weapon systems and key personnel. The plan includes—
  - Vehicle number, key leader, key weapon systems, additional personnel, and communications.
  - Location of PL.
  - Location(s) of PSG and medic.
  - Location of WSL.
  - Location(s) of communications (FO, RTO, or both).
- Ground movement plan:
  - Ensure troops are awake, alert, and pulling active security during movement.
  - Task the PL and vehicle commanders to track route progress.
  - Provide a compromise and contingency plan.
  - React to IEDs.
  - React to ambushes.
  - Respond to vehicle breakdowns.
- Unloading plan:
  - Dismount vehicles in accordance with the SOP and reverse loading plan.
  - Establish security.
  - Have the PSG account for personnel and clear all vehicles for departure.
  - Establish security at the halt or perimeter.
  - Adjust the perimeter appropriately as vehicles depart the area.
- Ground tactical plan:
  - Prepare to continue movement.
  - Conduct follow-on mission.

13-4. The WARNORD brings together the vehicle movement. It contains basic information on the situation, mission, task organization, any special instructions, initial time organization, and uniform and equipment common to all. (See table 13-1.)

Table 13-1. Mounted tactical movement brief

ADMINISTRATIVE. PERSONNEL (ROLL CALL).
1. **Responsibilities.**
   a. Driver / Nav.
   b. VCS drivers (primary and alternate).
   c. CSW operator.
   d. Counterassault element leader.
   e. Designated marksman or markswoman.
   f. Medic / Combat lifesaver.
   g. Guide / Interpreter.
   h. Higher HQ rep.
2. **Sectors of fire:** By priority, weapon system, vehicle, and phase.
3. **Task organization:** Internal organization for mounted patrol—manifest.

1. **SITUATION.**
   a. **Enemy forces:** Discuss enemy.
      (1) Identification of any known enemy.
      (2) Composition, capabilities, strengths, and equipment.
      (3) Location. (Highlight danger areas on map.)
      (4) Most likely and most dangerous course of action: DRAW-D.
   b. **Weather:** General forecast.
   c. **Light data:** EENT, percent illumination, moonrise, moonset, BMNT.
   d. **Friendly forces.**
      (1) Units along the route.
      (2) Operational support provided by higher HQ.
      (3) Aviation support.
         (a) ASOC   Call sign _[add here]_ Frequency _[add here]_
         (b) DASC   Call sign _[add here]_ Frequency _[add here]_
         (c) JSTARS Call sign _[add here]_ Frequency _[add here]_
      (4) Mobile security forces / Quick reaction forces.
         (a) EOD. SOF.
         (b) Fire support elements.
            _1_ Element _[add here]_ Location _[add here]_ Frequency / Call sign _[add here]_
         (c) Attachments. From outside the organization.
2. **MISSION.** Who, what, when, where, why (for example, Unit X conducts tactical mounted patrol to FOB YY and returns to FOB XX NLT 231000ZDEC03 in order to provide resupply of CL V [ammo]).

Table 13-1. Mounted tactical movement brief (continued)

3. **EXECUTION.**

    a. **Concept of operations:** Mounted patrol execution and tasks of elements, teams, and individuals at the objective(s). Use broad, general description beginning to end.

    b. **Tasks to subordinate units:** Include attached or OPCON elements.

    c. **Coordinating instructions:** Instructions for all units.

        (1) Safety. (See appendix E, Risk Management.)

            (a) Overall risk to force: Low   Medium   High   [Indicate one.]

            (b) Overall risk to mission accomplishment: Low   Medium   High   [Indicate one.]

            (c) Fratricide reduction measures.

        (2) Order of march: Spacing of serials; locations of support elements.

        (3) Routes: Attach strip map.

        (4) Additional movement issues: Speed, intervals, lane, parking, accidents, and other potential issues.

        (5) Timeline.

            (a) Vehicle and personnel gear preparation and completion of preventive maintenance checks and services.

            (b) Briefing.

            (c) Donning equipment.

            (d) Vehicle loading.

            (e) Rehearsals and test fire.

            (f) Backbrief and confirmation brief from key leaders.

            (g) Start point and departure.

            (h) Return to base.

            (i) Debriefing.

            (j) Recovery: Maintain vehicles and personnel gear.

        (6) Sectors of fire: Cover assigned sectors while mounted and dismounted. Cover up and down bridges, rooftops, balconies, storefronts, multistory structures, and crossroads.

        (7) Scanning: Scan crowds, vehicles, and roadsides for attack indicators. Note and communicate indicators throughout the mounted patrol.

            (a) Beware of motorcycles, vans with side doors, and dump trucks.

            (b) Beware of objects in the road (for example, parked cars, potholes, fresh asphalt or concrete, and trash).

        (8) Mounted patrol speed: [add here]   Min / Max: [add here] / [add here]

            (a) Either the rear vehicle's ability to keep up or the placement of the slowest vehicles in the lead dictates the patrol's speed.

            (b) Highways and open roads example: 50+ mi/h.

            (c) Urban and channeled areas: As fast as traffic allows. Brief evasive maneuvers, bumping and blocking technique, and ramming techniques that allow for continuous movement of the mounted patrol.

Table 13-1. Mounted tactical movement brief (continued)

(9) Vehicle interval.
    (a) Highways, open roads, cloverleaf interchanges, bridges, and ramps: Open spacing. Yet, vehicles are not to enter the mounted patrol.
    (b) Urban and channeled areas: Close interval. Maintain a visual of the tires on the vehicle in front of your vehicle even when this requires driving on the wrong side.
(10) Headlight status: On or off, blackout, or use of night observation.
(11) Rules of engagement for mounted patrol operations (theater-specific).
(12) Battle drills will be rehearsed: No need to cover in the brief.

4. **ADMINISTRATION AND LOGISTICS.** (Equipment) a. **Individual equipment:** See precombat inspection checklist. b. **Vehicles:** See precombat inspection checklist.
5. **MISSION COMMAND.** a. Chain of command: Positioning of mounted patrol. b. Mounted patrol call sign(s): _____[add here]_____ c. Area of operations communications and MEDEVAC and CASEVAC plans. d. Mounted patrol PACE communications. e. Vehicle internal (back to: _____[add here]_____). f. Hand and arm and visual signals: In accordance with the unit's standard operating procedure.
Legend: ammo–ammunition; ASOC–air support operations center; BMNT–begin morning nautical twilight; CASEVAC–casualty evacuation; CL V–Class V—ammunition and explosives; CSW–crew-served weapon; DASC–direct air support center; DEC–December; DRAW-D–defend, reinforce, attack, withdraw, and delay; EENT–end of evening nautical twilight; EOD–explosive ordnance disposal; FOB–forward operating base; HQ–headquarters; JSTARS–Joint Surveillance Target Attack Radar System; Max–maximum; MEDEVAC–medical evacuation; mi/h–miles per hour; Min–minimum; Nav–navigator; NLT–not later than; OPCON–operational control; PACE–primary, alternate, contingency, emergency; rep–representative; SOF–special operations forces; VCS–Vehicle Control System

13-5 Any involvement of a mounted patrol, especially in hostile environments, includes the possibility of an ambush, a forced stop, or another potentially hazardous situation. Rangers receive thorough training in maneuvers to enable them to protect themselves and their fellow Soldiers in these circumstances. Figure 13-1 on page 13-6 and figure 13-2 on page 13-7 detail various methods mounted patrols use to react to ambush.

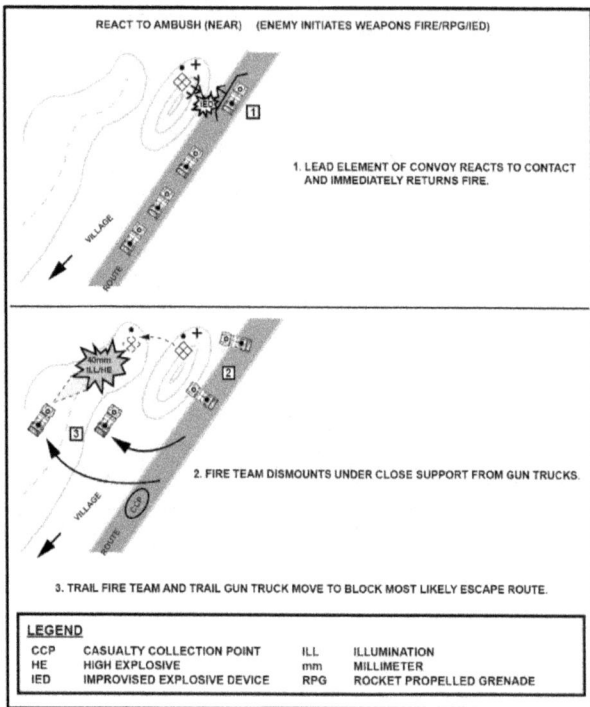

Figure 13-1. Mounted patrol reacting to near ambush

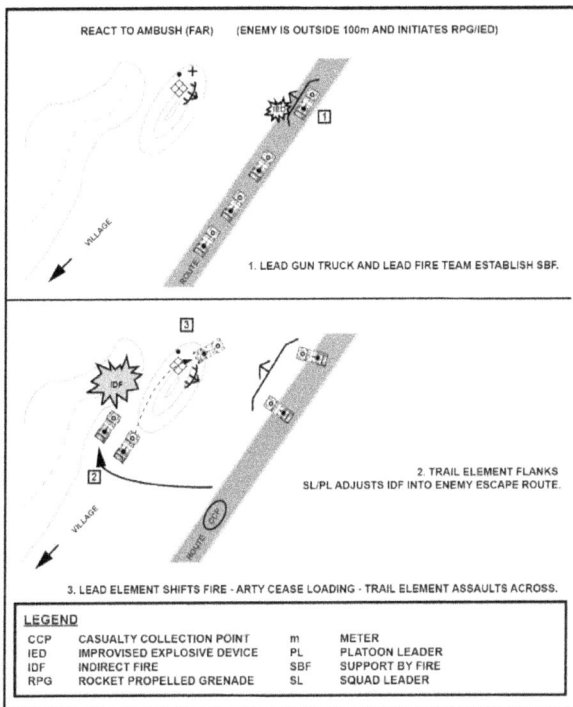

Figure 13-2. Mounted patrol reacting to far ambush

FORCED STOPS

13-6. When weapons fire, rocket-propelled grenades, IEDs, or indirect fire force(s) vehicles to stop, activate the turn signal to indicate the direction of contact. Once the vehicles are not in direct contact, report via internal communications the identity of the vehicle, type of contact, clock direction, and any available grid coordinates.

13-7. Personnel on vehicles forced to stop dismount on the noncontact side, assume covered positions, and provide an initial base of fire. The entire patrol halts, the remaining personnel dismount on the noncontact side, and they provide additional fire. Vehicles not in contact reposition and provide supporting fire.

METHOD ONE

13-8. The PL assesses the situation and maneuver in order to suppress the enemy and gain fire superiority. Once the PL determines the threat is eliminated, recovery and CASEVAC operations begin. When the PL determines the patrol cannot gain fire superiority to eliminate the threat, the patrol executes break contact procedures. Figure 13-3 details method one for mounted patrols to use when forced to stop.

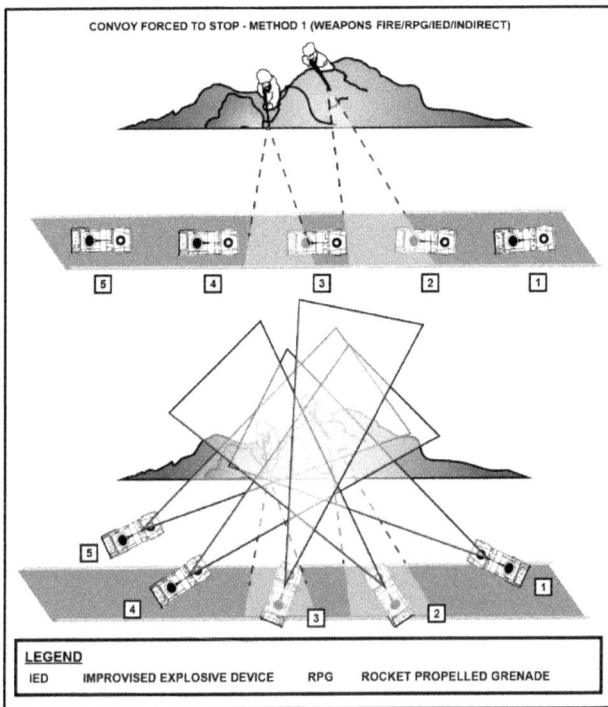

Figure 13-3. Mounted patrol forced to stop, method one

METHOD TWO

13-9. In this forced stop scenario, all personnel <u>stay in vehicles</u>. Drivers move the vehicles out of the kill zone. Any vehicles directly behind disabled vehicles push those disabled vehicles out of the kill zone. Operable vehicles establish a base of fire toward the suspected or known enemy.

13-10. Whenever the patrol gains fire superiority, the PL uses the minimum amount of force necessary to destroy the enemy. On the other hand, when the PL determines the patrol cannot gain fire superiority, the leader breaks contact. Figure 13-4 details method two for mounted patrols to use when forced to stop.

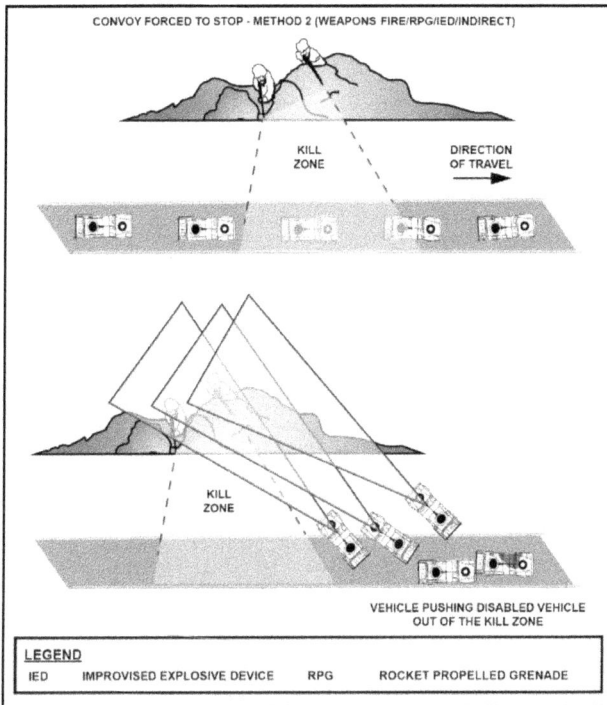

Figure 13-4. Mounted patrol forced to stop, method two

BREAK CONTACT

13-11. Always try to close with the enemy first to prevent their returning later to attack the patrol. When the PL determines the patrol cannot gain fire superiority and decides to break contact, the PL determines a rally point to the front, to the rear, or to both. Communications and pyrotechnic signals are useful for breaking contact and occupying rally points. The patrol deploys any available obscuration measures.

13-12. Using cover and concealment, aid and litter teams evacuate all casualties under fire. The patrol maintains position and fire suppression in the contact zone and assists aid and litter teams as necessary.

13-13. Leaders direct the towing or destruction of disabled vehicles. Operable vehicles displace forward or backward under the control of leaders. The most forward vehicle in the contact zone moves first, followed by the next most forward vehicle. Vehicles continue to displace under supporting fire until contact is broken.

13-14. If the breaking of contact occurs with vehicles on both sides of the contact zone, displacement of vehicles occurs using an alternating technique. Upon occupation of the objective rally point, leaders immediately position vehicles to establish 360-degree security, consolidate, and reorganize. Figure 13-5 details how to break contact.

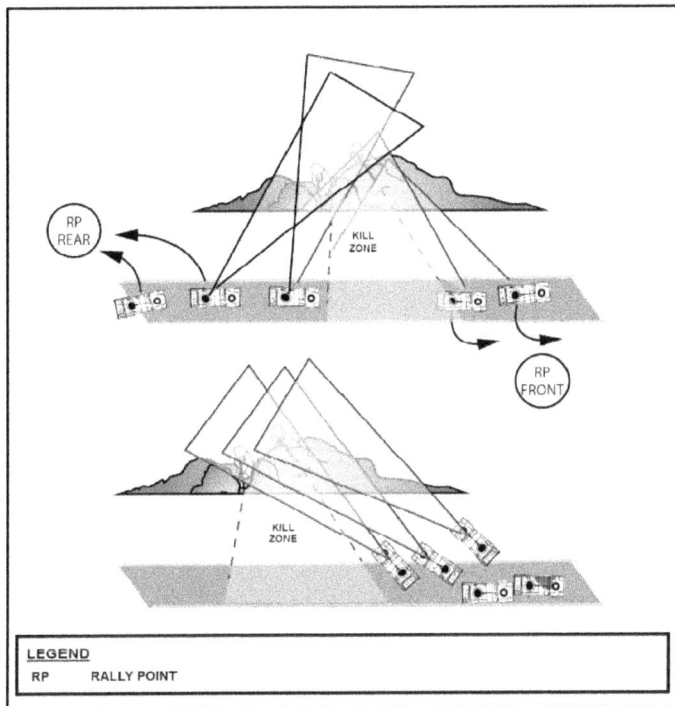

Figure 13-5. Mounted patrol breaking contact

CASUALTY EVACUATION AND RECOVERY OPERATIONS

13-15. Once the leader assesses the enemy threat is eliminated and the area is secure, CASEVAC and recovery operations begin. This helps focus Soldiers on defeating and destroying the threat. During CASEVAC operations, aid and litter teams position themselves on the safe side to extract the casualties and other personnel. The casualties receive treatment after their safe removal from the contact area.

13-16. During vehicle recovery procedures, recovery teams position themselves on the safe side of the disabled vehicle. The truck commander dismounts and assesses the disabled vehicle. Whenever a truck commander determines a vehicle can be recovered safely, the truck commander guides the recovery vehicle into position and conducts a hasty hookup. The truck commander remains ready and able to operate the disabled vehicle.

13-17. After exiting the contact area, the patrol conducts complete, correct hookup procedures. When the leader determines outside support is necessary for recovery, the leader contacts higher HQ for guidance. Leaders abandon or call for the destruction of disabled vehicles. Once recovery operations are complete, the team displaces and conducts linkup at the rally point. Figure 13-6 details CASEVAC and recovery operations.

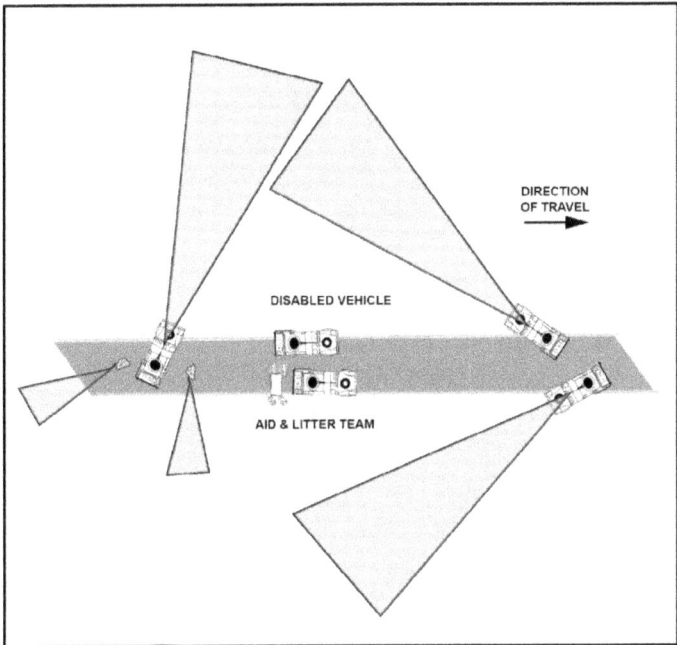

Figure 13-6. Casualty evacuation and recovery operations

This page intentionally left blank.

Chapter 14

Aviation

Leaders fully integrate Army aviation and Infantry units with other members of the combined arms team to form powerful and flexible air assault task forces. These forces project combat power throughout the depth and width of the modern battlespace with little regard for terrain barriers, thereby bringing about deliberate, precisely planned, and vigorously executed combat operations. They strike the enemy when and where they are most vulnerable.

REVERSE PLANNING SEQUENCE

14-1. The successful execution of an air assault originates with a careful analysis of the METT-TC (I) variables and detailed and precise reverse planning. Leaders develop five basic plans, which compose the reverse planning sequence, for each air assault operation. The battalion is the lowest level with sufficient personnel to plan, coordinate, and control air assault operations. For the conduct of company-sized or lower operations, most of the planning occurs at the battalion or higher HQ. The five plans are ground tactical, landing, air movement, loading, and staging.

14-2. The commander's ground tactical plan forms the foundation of a successful air assault operation. All additional plans support this plan. It specifies actions in the objective area to accomplish the mission and addresses subsequent operations.

14-3. The landing plan, in support of the ground tactical plan, outlines a sequence of events that allows elements to move into the AO. In this sequence, units arrive at designated locations and at prescribed times. They arrive prepared to execute the ground tactical plan.

14-4. The leader bases the air movement plan on the ground tactical and landing plans. It specifies the schedule and provides instructions for the air movement of troops, equipment, and supplies from PZs to LZs.

14-5. The leader bases the loading plan on the air movement plan. It arranges for the loading of troops, equipment, and supplies onto the correct aircraft. The planning of aircraft loads maintains unit integrity. Cross-loading is sometimes necessary to ensure survivability of mission command assets and to ensure the mix of weapons arriving at the LZ is ready to support the fight.

14-6. The leader bases the staging plan on the loading plan. It prescribes the arrival time of ground units including troops, equipment, and supplies at the PZ in the order of movement.

14-7. Small-unit leaders consider the size, surface conditions, ground slope, obstacles, and approach and departure when selecting a PZ or LZ. A minimum circular landing point separation from other aircraft and obstacles is necessary. The diameters of the PZs/LZs for specific aircraft follow:

- OH-6A = 25 meters (82 feet).
- AH-1 = 35 meters (115 feet).
- UH-60L and AH-64 = 50 meters (164 feet).
- Cargo helicopters = 80 meters (262 feet).
- Any helicopter with a sling load = 100 meters (328 feet).

14-8. Physical hazards such as sand, blowing dust, snow, tree stumps, or large rocks contaminate the surfaces of PZs and LZs and require resolving. Ground slope also affects landing and is therefore a concern. The acceptable degree ranges of slopes for landing follow:

- 0 to 6 percent for upslope.
- 7 to 15 percent for side slopes.
- Over 15 percent for no touchdown (when the aircraft can hover).

14-9. When planning the approaches and departures of the PZ and LZ, a useful obstacle clearance has a 10-to-1 ratio. For example, a 10-foot (3-meter) tree requires 100 feet of horizontal distance for approach or departure. Mark obstacles with a red chemical light stick at night or red panels in the day. Avoid using markings visible to the enemy.

14-10. Approach and depart facing into the wind and along the long axis of the PZ or LZ. The greater the load, the larger the PZ or LZ necessary to accommodate the insertion or extraction. The time of day or night requires different methods for marking the PZ and LZ. Here are examples for both lighting conditions:

- Day — A ground guide marks the PZ or LZ for the lead aircraft either by holding an M4 rifle overhead, displaying a folded VS-17 signal panel chest-high, or another coordinated and identifiable means.
- Night — The code letter **Y** or the inverted Y (see figure 14-1) marks the landing point of the lead aircraft at night. Chemical light sticks or bean bag lights serve under any light discipline. A swinging chemical light stick also serves to mark the landing point.

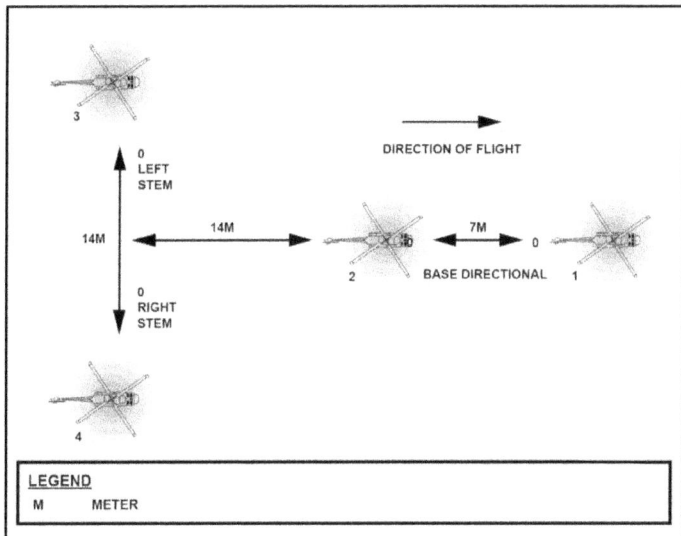

Figure 14-1. Inverted Y

AIR ASSAULT FORMATIONS

14-11. Aircraft supporting an operation uses any of the following PZ and LZ configurations. (See table 14-1.) The air assault task force commander, working with the air mission commander, prescribes these.

Table 14-1. Air assault formations

FORMATION	PROS	CONS
Heavy left or heavy right	Provides firepower to the front and flank.	Requires a relatively long, wide landing area. Presents difficulty in prepositioning loads. Restricts suppressive fire by inboard gunners.
Diamond	Allows rapid deployment for all-around security. Requires only a small landing area.	Presents some difficulty in prepositioning loads. Restricts suppressive fire of inboard gunners.
Vee	Allows rapid deployment of forces to the front. Requires a relatively small landing area.	Presents some difficulty in prepositioning loads.
Echelon left or echelon right	Allows rapid deployment of forces to the flank. Allows unrestricted suppressive fire by gunners.	Requires a relatively long, wide landing area. Presents some difficulty in prepositioning loads.
Trail	Allows rapid deployment of forces to the flank. Allows unrestricted suppressive fire by gunners. Requires a relatively small landing area. Simplifies prepositioning of loads.	Requires a long landing area.
Staggered trail left or right	Allows rapid deployment for all-around security. Simplifies prepositioning of loads.	Requires a relatively long, wide landing area. Somewhat restricts gunners' suppressive fire.

PICKUP ZONE OPERATIONS

14-12. Prior to the arrival of aircraft, the PZ is secured. The PZ control party is positioned, and the troops and equipment are positioned in the platoon and squad assembly areas. In occupying a patrol and squad assembly area, the patrol leader or SL maintains all-around security of the assembly area, maintains communications, organizes personnel and equipment into chalks and loads, and conducts safety briefings and equipment checks of the troops.

14-13. Figure 14-2 provides an example of a large, one-sided PZ. Figures 14-3 through 14-6 on pages 14-6 through 14-9 demonstrate loading and unloading procedures and techniques.

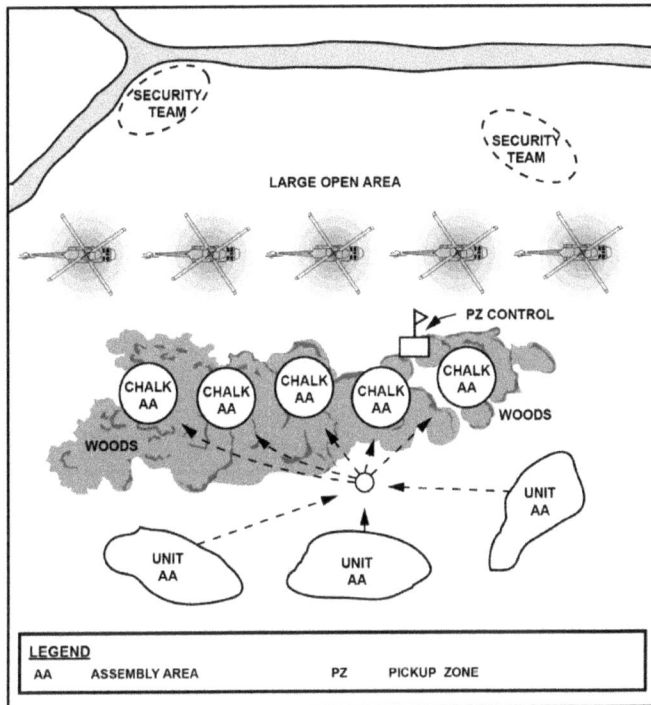

Figure 14-2. Example of large, one-sided pickup zone

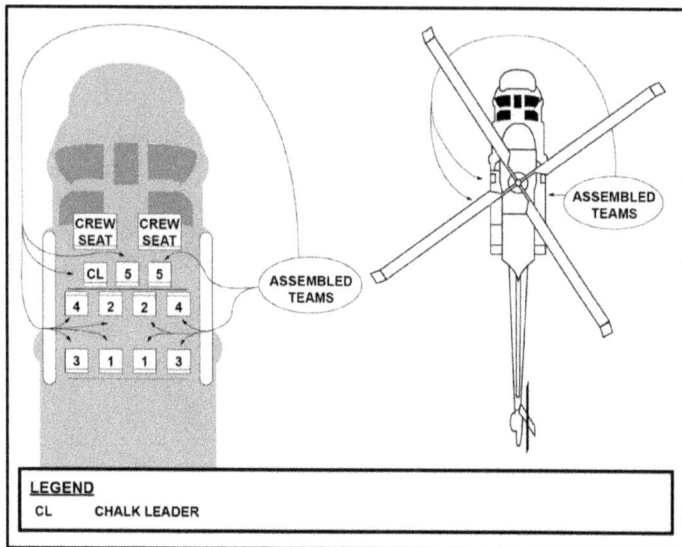

Figure 14-3. UH-60L loading sequence

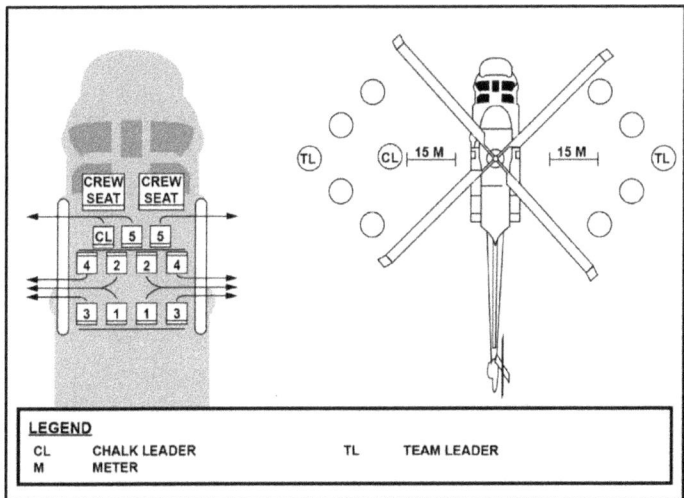

Figure 14-4. UH-60L unloading sequence

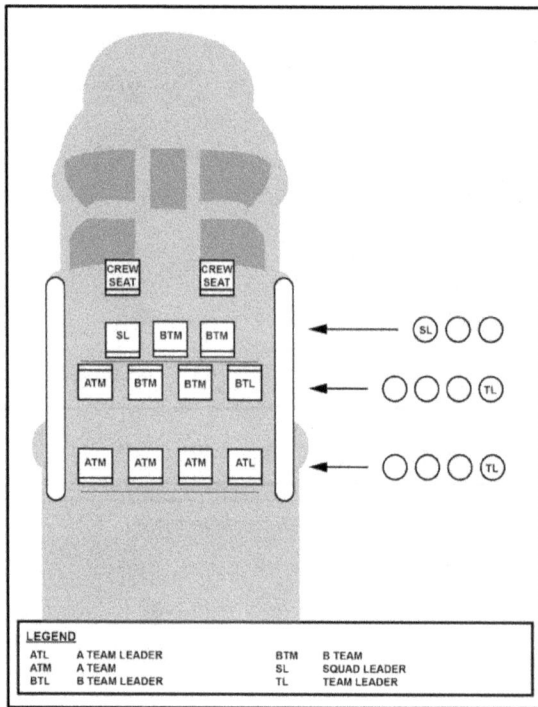

Figure 14-5. Tactical loading sequence

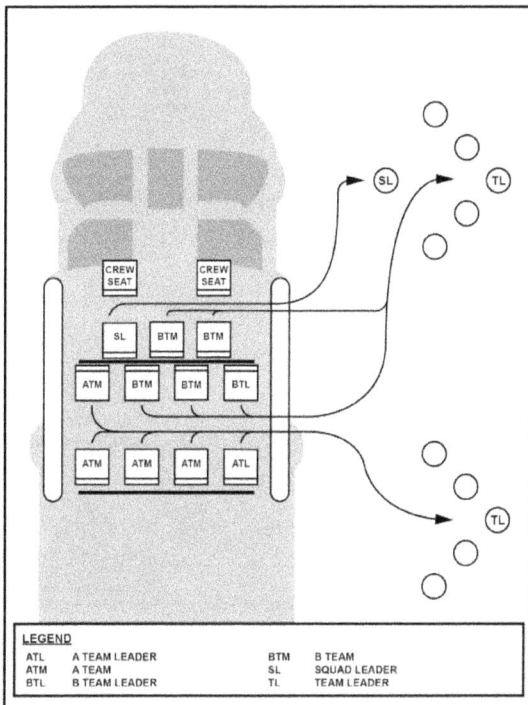

Figure 14-6. Tactical unloading sequence using door nearest cover and concealment

CH-47/CV-22 REAR RAMP OFF-LOAD

14-14. In the rear ramp off-load method, Soldiers exit from the rear ramp of a CH-47 or other rear-exiting aircraft. Soldiers move out from the aircraft and drop to a prone fighting position to establish 360-degree security until the aircraft lifts to depart the LZ. (See figure 14-7.) Once the aircraft departs the LZ, the unit executes a one- or two-side LZ rush in accordance with the landing plan or SOPs.

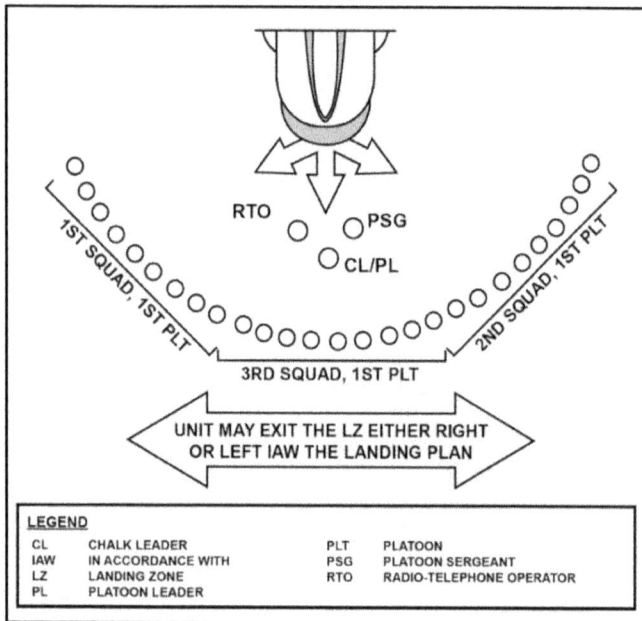

Figure 14-7. CH-47/CV-22 rear ramp off-load method

14-15. Safety is all leaders' primary concern when Rangers are operating in and around aircraft. The inclusion of aircraft into Ranger operations brings high risks. The following are some safety tips specifically related to the UH-60L.

- Approach the aircraft from 45 to 90 degrees off the nose.
- Point weapon muzzles upward when weapons are loaded with blank firing adapters.
- Point weapon muzzles downward when weapons are loaded with live ammunition.
- Wear a ballistic helmet.
- When possible, conduct an aircrew safety briefing with all personnel.
- At a minimum, cover loading, off-loading, emergency, and egress procedures.
- Leaders carry a manifest and turn in a copy to higher HQ.

REQUIREMENTS

14-16. Minimum landing space requirements and a minimum distance between helicopters on the ground depend on many factors. When the aviation unit's SOP does not spell out these requirements, the aviation unit commander works with the Pathfinder leader. The final decision about minimum landing requirements rests with the aviation unit commander. In selecting helicopter landing zones (HLZs) from maps, aerial photographs, and actual ground or aerial reconnaissance, consider the following factors:

- Number of helicopters — To land a large number of helicopters at the same time, the commander sometimes provides additional HLZs nearby. Alternatively, the commander has the helicopters land at the same HLZ but in successive lifts.
- Landing formations — Helicopter pilots match the landing formation to the flight formation whenever possible. Pilots modify their formations to accommodate the restrictions of an HLZ no more than necessary. However, landing in a restrictive area sometimes requires they modify their formations.
- Surface conditions — Rangers choose HLZs with firm surfaces. These prevent helicopters from bogging down, creating excessive dust, or blowing snow. Rotor wash stirs up any loose dirt, sand, or snow, which obscures the ground and especially so at night. Rangers remove airborne and any other debris from landing points to avoid damage to the rotor blades and turbine engines.
- Ground slope — Rangers choose HLZs with relatively level ground. A slope exceeding 7 degrees jeopardizes the safe landing of a helicopter. Whenever possible, pilots land upslope rather than downslope. All helicopters can land where ground slope measures 7 degrees or less.

14-17. Day operation signals for daylight operations use a different smoke color for each HLZ. Use of the same color more than once is possible when its distribution is spread out. The use of smoke is only as necessary—since it is visible to the enemy—and only when a pilot asks for help locating the HLZ.

14-18. Night operation signals for nighttime operations use pyrotechnics or other visual signals in lieu of smoke. As in daylight, red signals mean DO NOT LAND, but they also serve to indicate other emergency conditions. Everyone plans and knows the emergency codes. Each flight lands at the assigned site in accordance with the messages and the displayed visual aids. Hand and arm signals are useful for controlling the landing, hovering, and parking of helicopters.

PLANNING CONSIDERATIONS

14-19. To ensure success of the ground mission, leaders plan their own missions in detail. The more time they have to make plans, the more detailed of plans they can make. As soon as the senior leader receives word of a pending operation, a mission alert is issued and immediately followed with a WARNORD. These contain just enough information for the subordinate leaders to start preparing for the operation. These include the following:

- Roll call.
- Enemy and friendly situations (in the brief).
- Mission.
- Chain of command and task organization.
- Individual uniform and equipment (if not discussed in the SOP).
- Requisite equipment.
- Work priorities including who does what, when, and where.
- Specific instructions.
- Attached personnel.
- Coordination times.

14-20. Upon receiving the alert or WARNORD, leaders inspect and appropriately augment personnel and equipment. Leaders prepare equipment in the following order, from the most to the least important:

- Radios.
- Navigation aids (electronic and visual).
- Weapons.
- Essential individual equipment.
- Assembly aids.
- Other necessary items.

14-21. To succeed, an operation requires security. Each person receives only the information necessary to complete each phase of the operation. For example, the commander isolates any Soldiers who know the details of the operation. The situation dictates this extent of security requirements.

ROTARY-WING AIRCRAFT SPECIFICATIONS

14-22. Rotary-wing aircraft are vital for the success of certain missions. The Army relies on different types such as the UH-60L (see table 14-2), CV-22 (see table 14-3), and the CH-47D (see table 14-4 on page 14-14).

14-23. When fitted with a sling load, the CH-47D technical data package is #5 (100-meter diameter). Without the sling load, it is #4 (80-meter diameter). Specifications for all three helicopters are in the following tables.

Table 14-2. UH-60L specifications

Optics	Pilots use AN/AVS-6 to fly the aircraft at night.
Navigation Equipment	Doppler navigation set or Global Positioning System
Flight Characteristics	Maximum speed (level): 156 knots Normal cruise speed: 120 to 145 knots Maximum speed (with external sling loads): 90 knots
Additional Capabilities	The External Stores Support System allows for extended operations (longer than 5 hours) without refueling the two 230-gallon (871-liter) fuel tanks. The System also allows configuration for ferry and self-deployment flights with four 230-gallon (871-liter) fuel tanks. For enhanced mission command, the console provides the maneuver commander with an airborne platform capable of supporting two very high frequency radios, six secure very high frequency radios, one high frequency radio, and two ultrahigh frequency radios. The UH-60L is capable of inserting and extracting troops with the Fast-Rope Insertion and Extraction System and the Special Patrol Insertion and Extraction System. Sling load lift rating: 9,000 pounds (4,082 kilograms)

Table 14-3. CV-22 specifications

Weapon System and Range	M2 (Caliber .50 machine gun): 1,800 meters	
Communications Equipment	Internal	AN/AIC-30
	External	ARC-210 radio
Navigation Equipment	Navigational aid	ARN-147
Flight Characteristics	Cruise airspeed: 240 knots	Maximum carrying capacity: 24 seated troops, 32 floor-loaded troops, or 10,000 pounds (4,536 kilograms)
	Maximum airspeed: 240 knots	Endurance: 500 nautical miles with troops
Aircraft Survivability Equipment	Radar warning receiver	AN/APR-39A(V)2
	Laser warning	AN/AVR-2A Laser Detection System
	Missile warning	AN/AAR-47
	Electronic countermeasures	ALE-47 Countermeasures Dispensing System
Fuel Capacity	2,025 gallons (7,665 liters)	
Other Capabilities	Self-deployable, vertical, or short takeoff and landing	

Table 14-4. CH-47D specifications

Optics	Pilots use AN/AVS-6 to fly the aircraft at night.
Navigation Equipment	Doppler navigation set or Global Positioning System
Flight Characteristics	Maximum speed (level): 170 knots Normal cruise speed: 130 knots Maximum carrying capacity: 33 to 55 troops, 24 litters and 3 attendants, or 28,000 pounds (12,701 kilograms)
Additional Capabilities	One alternate configuration allows for additional fuel for mobile forward-aiming refueling equipment system or for ferry and self-deployment missions. The CH-47D has an internal load winch to ease the loading of properly configured cargo. The CH-47D can sling-load virtually any piece of equipment in the light Infantry, Airborne, or air assault divisions.

Chapter 15

Tactical Combat Casualty Care

Patrolling, more than some other types of missions, puts Rangers in harm's way. Trained medical personnel are not necessarily available at the initial point of injury. For this reason, Rangers understand CCP procedures and casualty triage, treat injuries by the principles of tactical combat casualty care, and understand the best methods of evacuation for moving Ranger casualties to higher levels of care.

GOALS AND PHASES

15-1. Tactical combat casualty care focuses on identifying and treating the most common causes of preventable death in the battlespace: hemorrhage/bleeding, tension pneumothorax, and airway issues. The three main goals of tactical combat casualty care are to treat the casualty, prevent additional casualties, and enable the unit to complete the mission.

15-2. Tactical combat casualty care is organized into phases of care, which start at the point of injury. These phases are relevant to combat and noncombat trauma scenarios. Paragraphs 15-3 through 15-5 describe the three phases of care.

15-3. Care under fire, or care under threat, is the aid rendered at the trauma scene while the threat is still active. Available medical equipment is limited to that carried by an individual or found in a nearby first aid kit. Massive bleeding is the only medical priority requiring attention during this phase due to the ongoing threat in a potentially chaotic and dangerous situation.

15-4. Tactical field care (TFC) is the care provided after the neutralization of the threat when the scene is safe or the casualty is no longer in the immediate threat situation. The performance of a rapid casualty assessment occurs during this phase. Bleeding control is assessed or reassessed as appropriate. Airway and breathing issues are addressed. Other injuries such as burns, fractures, eye trauma, and head injuries are now identified and treated. Medical equipment is most likely still limited. The time of the arrival of medical personnel or of evacuation varies considerably depending on the tactical situation. This care is usually associated with treatment at a CCP.

15-5. Tactical evacuation care is the care rendered during and once an aircraft, vehicle, or other mode of transportation for evacuation has moved the casualty to a higher level of care. Additional medical personnel and equipment are typically available in this phase of casualty care.

CARE UNDER FIRE

15-6. The mission does not stop for a casualty. Most battlespace casualty scenarios involve rapid medical and tactical decision making. The combat environment affords no stoppage when casualties occur. Good medicine is sometimes bad tactics; doing the right thing at the wrong time can result in additional casualties or mission failure. The tactical situation dictates the order of initial actions. Little time is available to provide casualty care while the unit is under effective enemy fire. Suppressing hostile fire and gaining fire superiority remain the priorities to minimize the risk of injury to additional personnel, as well as the risk of additional injury to the casualty, while completing the mission. Personnel sometimes assist in returning fire instead of stopping to care for casualties; this includes the casualty who is still able to fight. Leaders direct wounded service members exposed to enemy fire to continue returning fire, to move as quickly as possible to any nearby cover, and to perform self-aid.

15-7. In care under fire, the utmost priorities are to discover and stop massive bleeding and to get the casualty and oneself to cover and out of hostile fire. A pulsing or steady bleeding from a wound or traumatic amputation of an extremity indicate massive bleeding to a caretaker. Upon identifying massive bleeding, the caretaker applies a tourniquet without delay. Injury to a major vein or artery can result in shock or death from blood loss in minutes. An extremity (arm or leg) hemorrhage is a leading cause of preventable combat death. The use of one or more tourniquet(s) to stop the bleeding is essential to the survival of casualties with these types of injures.

TACTICAL FIELD CARE

15-8. TFC is the care rendered by a first responder or CLS once the responder and casualty are no longer under direct threat from effective enemy fire. This affords the time and relative safety for a more deliberate approach to casualty assessment and treatment.

15-9. Casualty assessment and management in TFC. Follow the MARCH PAWS approach:

- During threat to life:
  - Massive bleeding — Control with at least one tourniquet.
  - Airway — Control by opening or with the use of an airway adjunct.
  - Respiration (breathing) — Treat all open and sucking chest wounds with a vented chest seal.
  - Circulation — Reassess all tourniquets and treat for shock.
  - Hypothermia / Head injuries — Dry and warm the patient and prevent further injury to a suspected head injury.
- After threat to life:
  - Pain — Manage with pain medication found in the joint first aid kit.
  - Antibiotics — Use those found in the joint first aid kit.
  - Wounds — Dress any additional wounds.
  - Splints — Splint any fracture.

Note. These are helpful mnemonic acronyms for remembering how to approach casualty assessment and management systematically and to identify and treat life-threatening injuries promptly, thereby saving lives in the battlespace and reducing preventable combat deaths.

MASSIVE BLEEDING

15-10. Quickly identify and control bleeding. Apply a tourniquet to arterial bleeding of the extremities, high and tight and over the uniform. When this does not control the bleeding, apply a second tourniquet above the first. When tourniquets cannot control the bleeding due to the location of the wound, use combat gauze. Control all other bleeding with a pressure dressing. Check dressings often to ensure the bleeding is under control.

AIRWAY

15-11. Airway obstruction, or a blockage in the airway, usually occurs at the base of the tongue. When this is the case and there are no traumatic injuries to the face or skull, open the airway using the chin-lift method. When traumatic injuries exist, keep the airway open with a jaw thrust.

RESPIRATION/BREATHING

15-12. If the patient is having trouble breathing, expose the chest and identify open chest injuries. Check for entrance and exit wounds. Apply a vented chest seal to cover open entry and exit chest wounds. Place the patient on the injured side or in a position that allows more comfortable breathing. For suspected tension pneumothorax, burp the chest seal. When this does not seem to relieve the tension pneumothorax, refer to a medic.

CIRCULATION

15-13. Reassess all tourniquets and stop all other previously untreated bleeding. An inadequate flow of oxygen to body tissues causes shock. The most common form of shock is hemorrhagic, that which is due to uncontrolled bleeding. Signs and symptoms of shock include sweaty but cool skin, which is often called clammy; pale skin; restlessness; nervousness; agitation; unusual thirst; altered mental status; rapid breathing; blotchy, bluish skin around the mouth; and nausea. Basic treatment steps for shock are to—

- Control the bleeding.
- Open the airway.
- Restore breathing.
- Position the casualty.
- Monitor the casualty's condition.
- Evacuate the casualty.

HYPOTHERMIA/HEAD INJURIES

15-14. Minimize the casualty's exposure to the elements. Keep protective gear on or with the casualty as feasible. Replace wet clothing with dry whenever possible. Whenever the self-heating blanket from the hypothermia prevention and management kit is available, apply it to the torso along with the blizzard survival blanket. When these specific items are unavailable, use blankets, ponchos, sleeping bags, or anything that retains heat and keeps the casualty dry. Constantly monitor any casualty with a suspected head injury for declining mental status or the need for further evaluation.

PAIN

15-15. Pain treatment occurs once the unit has managed all life-threatening issues. When conscious, casualties can self-manage their pain with their combat wound medication packs found in their joint first aid kits, or medical personnel can administer injectable pain medications.

ANTIBIOTICS

15-16. The administration of antibiotics occurs once the unit has managed all life-threatening issues. The casualty can receive antibiotics in the same manner as pain management. The combat wound medication pack contains oral antibiotics the casualty can take when conscious, or medical personnel can administer injectable antibiotics.

WOUNDS

15-17. The treatment of wounds occurs once the unit has managed all life-threatening issues. Dress all known wounds and then assess all the applied bandages for increased pain, skin discoloration, and irregular pulse. These symptoms are indicative of an emergency or the need to loosen the bandages. Any loosening of bandages requires fresh monitoring for excessive bleeding.

SPLINTS

15-18. The application of splints occurs once the unit has managed all life-threatening issues. The splinting of fractures or otherwise unstable extremities has the capabilities of relieving the casualty's pain and protecting the extremity from further injury.

CASUALTY COLLECTION POINTS

15-19. During TFC, the mission and number of wounded sometimes render a CCP necessary. CCPs require security to protect both the wounded and those treating the wounded. The wounded require good cover and concealment to prevent further injuries. The best location of the CCP is close enough to the wounded that their treatment suffers no delay, but far enough from any effective fire that no additional casualties are added to their number. Again, the best placement of CCPs is near HLZs or ambulance exchange points to aid in rapid transport. Furthermore, the best positioning of the immediate casualties is closest to the HLZ or ambulance exchange point. The CCP location needs identifying prior to the actions on the objective, but leaders also prepare to do hasty CCPs while en route to the objective (en route rally points).

15-20. The PSG maintains responsibility of the CCP, which has strict command and control to enable casualty accountability and prompt treatment. The PSG assigns the appropriate medic and aid and litter teams or Rangers qualified in tactical combat casualty care to conduct triage, treatment, and transportation of the wounded. The establishing and equipping of aid and litter teams occurs before an operation commences. Designated aid and litter teams are aware of the location of the CCPs, ambulance exchange points, and medical treatment facilities and are capable of navigating to them as necessary. Aid and litter teams receive updates on changes to the locations of CCPs, ambulance exchange points, or medical treatment facilities as soon as possible to avoid unnecessary delay in moving casualties and unnecessary exposure of personnel to additional hazards. Aid and litter teams are familiar with unit medical tactics, techniques, and procedures for CCPs including considerations for separate areas for those KIA, contaminated casualties, detainees, and civilians. Enemy casualties and the disposition of enemy equipment require a security plan. The PSG may request additional assets to assist in the treatment of the wounded based on the number and types of casualties.

15-21. The most medically qualified person triages all patients, helps instruct others in what treatments need completing, supervises the proper treatments of casualties, or performs lifesaving procedures. Sensitive items such as weapons, maps, overlays, and communications equipment are considered unit equipment and will be secured and returned to the unit.

Note. Aid and litter teams are a task-organized fire team whenever the situation permits. This provides internal leaders and communications and enables the team to react to contact were enemy action to interfere with casualty treatment.

15-22. The underline triangle method is one of many ways to set up and maintain a CCP (see figure 15-1 on page 15-6 for a depiction of this triangular organization of triage). This method is as follows:
- The tip of the triangle is the entrance and exit control point and the staging point for the medic or PSG to direct traffic.
- TFC occurs at the CCP.
- The triangle's left side (10 to 6 o'clock) is for the immediate patients with their heads facing inward.
- The triangle's right side (2 to 6 o'clock) is for the delayed patients with their heads facing inward.
- The triangle's base (10 to 2 o'clock) is for the minimal patients with their heads facing inward.

Note. Minimal patients can still pull security while undergoing evaluation for their injuries.

- Expectant patients and those KIA lie away from others in the CCP but are not alone whenever possible.
- Depending on the tactical situation, the use of chemical light sticks to mark the legs of the CCP is permissible during hours of limited visibility.

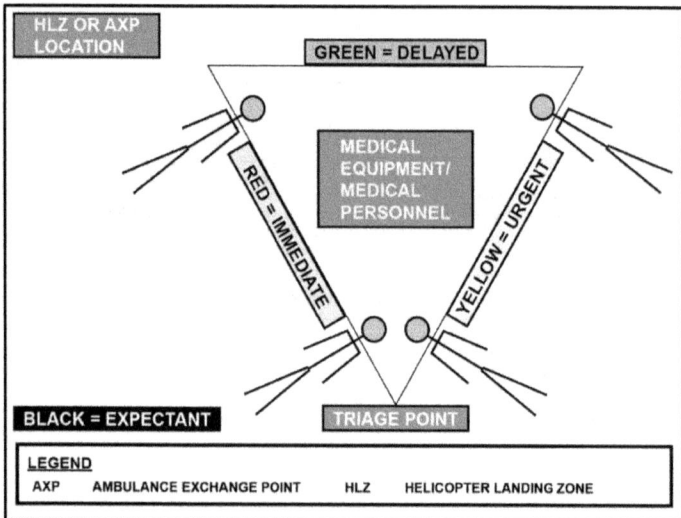

Figure 15-1. Example of triage triangle

15-23. Patient triage holds a critical place and requires continuous reevaluation for the entire time patients remain on the ground. The triage categories follow:

- Immediate — any patient who needs an immediate higher echelon of medical care (most likely requiring surgery) to life, limb, or eyesight within 1 to 2 hour(s).
- Delayed — any patient who needs a higher echelon of medical care but whose injuries an onsite medic can maintain in the presence of more urgent patients.
- Minimal — any patient who does not require a higher echelon of medical care and remains capable of carrying on with the mission if so directed (for example, broken hand, cut on leg).
- Expectant — patients whose wounds are so extreme as to render their survival unlikely even if they were the only patients and had all the available medical resources dedicated to their treatment and care.

Expectant patients remain separate from the others, kept as comfortable as possible, and are not abandoned.

15-24. A Ranger considers the following greater priorities, still utilizing MARCH (see paragraph 15-9), to decide how to prioritize the treatment of multiple casualties in the TFC phase:

- Massive bleeding.
- Penetrating trauma to the torso.
- Airway.
- Respiratory distress.
- Altered mental status.

TACTICAL EVACUATION CARE

15-25. Tactical evacuation care is the care rendered once an aircraft, ground vehicle, or other evacuation platform has picked up the casualty. This phase of care continues the care for the casualty that began in an earlier phase. Tactical evacuation care is similar to TFC in many ways. However, the extra medical personnel and equipment on the evacuation asset provide additional care in this phase of casualty care management.

15-26. The concept of tactical evacuation encompasses CASEVAC and MEDEVAC. Nonmedical first responders or CLSs are not expected to care for casualties during evacuation. However, were this to happen, the approach (MARCH PAWS) and skills used for care under fire and TFC apply.

15-27. The key principle in tactical evacuation care is that monitoring with appropriate assessment and treatment continues until medical personnel assume the care of the casualty or until the casualty arrives at a higher echelon of care. Pre-evacuation procedures include the first responder or CLS documenting all the assessments and care rendered for care under fire and TFC on DD Form 1380 (*Tactical Casualty Combat Care [TCCC] Card*).

15-28. An evacuation request uses the 9-line MEDEVAC format per a unit's SOPs to initiate a MEDEVAC. The MEDEVAC request sometimes includes a mechanism of injury, injury type, signs, treatment (MIST) report. Before evacuation, casualties are made ready for evacuation with their items such as weapons and equipment secured and with litter and evacuation equipment prepared.

15-29. The performance of a CASEVAC is by any other means than a medical platform. One alternative involves the Ranger unit's moving the wounded Ranger to further medical assets by using manual carries or a rescue and transport system. A CASEVAC that employs aid and litter teams is physically demanding for those team members. Leaders understand the physical demands and swap out litter team members as necessary in the sustainment of the CASEVAC support. The same is true for situations that involve manual carries. The tactical situation, limited number of rescue and transport systems, and the casualty's ability to move with assistance sometimes require manual carries. Leaders choose the best means of performing a CASEVAC by assessing the tactical situation. CASEVAC vehicles can be equipped with crew-served weapons.

15-30. A medical platform with dedicated medical personnel and potentially more medical equipment performs a MEDEVAC whether by land, sea, or air. MEDEVAC vehicles are properly marked and afforded protections in accordance with the Geneva Conventions and the law of war. Due to these laws, MEDEVAC ambulances may not be equipped with crew-served weapons. The Ranger unit is proficient in performing a 9-line MEDEVAC request and setting up an HLZ.

15-31. There are several limitations to using CASEVAC instead of MEDEVAC. Some such limitations to CASEVAC follow:

- A CASEVAC does not provide en route care unless preplanned and staffed.
- The CASEVAC vehicle lacks the medical equipment suite typical of a MEDEVAC ambulance.
- Lack of space often prevents an accompanying CLS or combat medic from performing lifesaving interventions and even impedes monitoring.
- A CASEVAC vehicle is not a dedicated evacuation platform and has competing requirements (for example, the movement of ammunition or water). As a result, CASEVAC vehicles are often not readily available but committed elsewhere.
- The CASEVAC vehicle lacks the protections afforded to properly marked MEDEVAC ambulances in accordance with Geneva Convention I and the law of war.

15-32. The two major methods Rangers use to mark HLZs for MEDEVACs are commonly called NATO T and inverted Y. These names refer to the shapes the markers make on the ground.

15-33. To lay out the NATO T marking, Rangers place a marker that establishes their middle, or center, point. Then, they move 10 meters to their left and place a marker. They return to the center point, go 10 meters to their right, and place a marker. They return to the center point once more, move down 10 meters, and place a marker. Finally, they move down 10 more meters and place their final marker. Figure 15-2 provides a visual depiction of the resulting marker placements that altogether resemble the capital letter **T**.

15-34. To lay out the inverted Y marking, Rangers establish their middle, or center, point. They move 7 meters to their left and place a marker. They return to the center point, move 7 meters to their right, and place a marker. They return to the center point once more, move up 14 meters, and place a marker. Finally, they move up 7 more meters and place their final marker. Figure 15-2 provides a visual depiction of the resulting marker placements that altogether resemble the capital letter **Y** on its head.

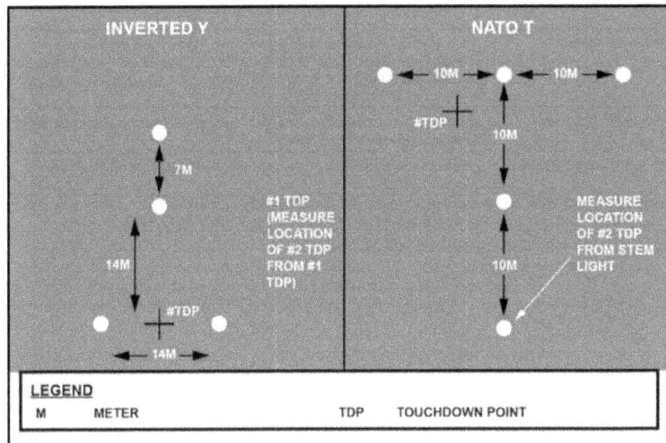

Figure 15-2. Inverted Y and NATO T layouts

MEDICAL EVACUATION REQUEST

15-35. The institutionalization of the procedures for requesting a MEDEVAC occurs down to the unit level. Table 15-1 provides procedural guidance and standards for MEDEVAC requests. Requests for both ground and aeromedical evacuation follow the same format. Rangers order MEDEVACs with a 9-line MEDEVAC request, and transmission of a 9-line MEDEVAC request always travels by secure means. (See appendix B for sample formats of both a 9-line MEDEVAC request and a MIST report.)

Table 15-1. Categories of evacuation precedence

Priority I—URGENT	Emergency cases requiring evacuation as soon as possible and within a maximum of 1 hour to save life, limb, or eyesight, to prevent complications of serious illness, or to avoid permanent disability.
Priority IA—URGENT-SURG	Patients requiring both (1) evacuation as soon as possible and within a maximum of 1 hour and (2) far-forward surgical intervention to save life, limb, or eyesight or to stabilize for further evacuation.

Table 15-1. Categories of evacuation precedence (continued)

Priority II—PRIORITY	Sick or wounded personnel requiring prompt medical care. The patient requires evacuation within 4 hours, or the patient's medical condition could deteriorate to such a degree to warrant URGENT precedence, requires special treatment not available locally, or could cause unnecessary pain or disability.
Priority III—ROUTINE	Sick or wounded personnel warranting evacuation within 24 hours but whose condition is not expected to deteriorate significantly.
Priority IV—CONVENIENCE	Patients for whom evacuation by medical vehicle is a matter of medical convenience rather than necessity.
Note: STANAG 3204 deleted category Priority IV—CONVENIENCE. However, U.S. Army evacuation priorities still include this category due to its ongoing necessity in an operational environment.	
Legend: SURG–surgery; U.S.–United States	

15-36. U.S. Soldiers regularly operate as part of a North Atlantic Treaty Organization (NATO) allied joint task force. During these operations, Soldiers receive or transmit 9-line MEDEVAC requests to and from NATO allies. (See ATP 4-02.2 and STANAG 2546 for information on a MEDEVAC request.) To avoid confusion, Soldiers understand the differences between the NATO and U.S. MEDEVAC request formats. The NATO MEDEVAC request format differs from the U.S. format in several ways:

- NATO does not use the urgent surgical category. The United States uses the urgent surgical category to ensure the direct evacuation of a patient requiring surgery to the closest surgical asset. This enables roll bypass when the closest medical treatment facility (such as a Role 2) does not have the capability for treatment that another facility further away (such as a Role 2 with a forward surgical detachment or a Role 3) does have.
- Line 6 on the NATO request only ever requires information about the security at the pickup site. The information it requests does change between two possible responses depending on the situation like the U.S. request does. On the U.S. request, when it is wartime, line 6 requests information on the security at the pickup site, but in peacetime, it requests information on the number and types of wounds, injuries, or illnesses.
- Line 8 on the NATO request has similar but different patient nationality status codes and definitions. The primary distinction lies in the U.S. request's specifying U.S. military or U.S. civilian status, while the NATO request uses simply military or civilian status. This difference can cause confusion when patient-regulating policies require specific facilities for civilians or by nationalities.
- Some multinational partners request additional information in conjunction with a 9-line MEDEVAC request. The inclusion of this additional information is possible by incorporating the MIST report into the 9-line MEDEVAC request.

15-37. The task steps for initiating a 9-line MEDEVAC request follow. The person making the request decides up front whether to append a MIST report in order to prepare that concurrently. Neither the 9-line MEDEVAC request nor the MIST report are formal, standardized forms, but they do follow a widely recognized format that recipients expect more likely than not. Following the format leads to the swiftest, clearest communication and the most accurate understanding on the receiving end, so responders arrive informed and prepared, with the proper expectations.

a. Collect all the necessary, applicable information for the MEDEVAC request (and for any MIST report to be included).
   (1) Determine the grid coordinates for the pickup site.
   (2) Obtain the receiving unit's radio frequency, call sign, and suffix.
   (3) Obtain the number of patients in each evacuation precedence category.
   (4) Determine the requisite special equipment.
   (5) Determine whether the patients are ambulatory or require a litter and the count of each type
   (2) Identify whether any patient requires escorting.
   (3) Determine the security at the pickup site.
   (4) Determine the marking(s) of the pickup site.
   (5) Determine the nationality and status of each individual patient and group them into the applicable types for a count of each category.
   (6) Obtain the pickup site's CBRN contamination information, which is normally available from senior personnel or medics.

Note. In wartime, line 9 on the 9-line MEDEVAC request provides information on CBRN contamination conditions. However, such information is included only when contamination exists.

b. Organize the gathered MEDEVAC information into the request format using the authorized brevity codes.
   (1) Location of the pickup site (line 1).
   (2) Radio frequency, call sign, and suffix (line 2).
   (3) Number of patients by precedence (line 3).
   (4) Special equipment required (line 4).
   (5) Number of patients by type (line 5).
   (6) Security at the pickup site (line 6).
   (7) Method of marking the pickup site (line 7).
   (8) Number of patients' nationalities and statuses by type (line 8).
   (9) CBRN contamination (line 9) when existent.
c. Transmit the MEDEVAC request.

Note. Transmission varies depending on individual experience levels and the situation.

   (1) Contact the unit who controls the evacuation assets.
      (a) Make proper contact with the intended receiver.
      (b) Use the effective call sign and frequency assignments from the signal operating instructions.
      (c) Give, in the clear, I HAVE A MEDEVAC REQUEST. Wait 1 to 3 second(s) for a response. Repeat the statement for no response.
   (2) Transmit the MEDEVAC information in the proper sequence.

(a) State all the line item numbers in clear text. Transmit any necessary call sign and suffix in line 2 in the clear.

Note. Always transmit the information for lines 1 through 5 during the initial contact with the evacuation unit. Transmission of the information for lines 6 through 9 sometimes transpires while the evacuation aircraft or vehicle is en route.

(b) Follow the procedure provided in the explanation column of the MEDEVAC request format to transmit other required information.
(c) Pronounce letters and numbers in accordance with the appropriate radio-telephone procedures.
(d) End the transmission by stating OVER.
(e) Keep the radio on and listen for additional instructions or contact from the evacuation unit.

Example in a Non-CBRN-contaminated Environment

"Ranger 61, this is Ranger 31, I have a MEDEVAC request, OVER."

"Ranger 31, this is Ranger 61, send your MEDEVAC request, OVER."

"LINE ONE, 17 PAPA UNIFORM 12345678,

LINE TWO, 35000, RANGER 31,

LINE THREE, ONE ALPHA,

LINE FOUR, ALPHA,

LINE FIVE, ONE LIMA,

BREAK,

LINE SIX, NOVEMBER,

LINE SEVEN, ALPHA,

LINE EIGHT, ONE ALPHA, OVER."

MASS CASUALTY INCIDENT

15-38. A mass casualty incident, by definition, is having more patients than the organic medical assets can handle. The most important actions in a mass casualty are to triage patients appropriately and to establish and manage the CCP. Leaders designate aid and litter teams before the mission. Aid and litter teams comprise members of the same fire team or squad to ensure proper command and control. Command and control during a mass casualty is paramount for ensuring emplaced security, accountability of the wounded, and complete and proper treatment. The Ranger unit sometimes makes, prior to their patrol, bump cards with their battle roster numbers for each unit member to carry in the same location so leaders can easily maintain accountability of the wounded and evacuated. The PL and PSG maintain communication with higher HQ to evacuate the wounded in a timely manner and receive further guidance on future missions.

FIELD SANITATION

15-39. Rangers sanitize their intended potable water with the iodine purification process. The procedure for using iodine to ensure potable water for drinking, cooking, and brushing one's teeth follows.
    a. Fill the canteen with the cleanest, clearest water available.
    b. Add two iodine tablets to each 1-quart canteen of water or four tablets to each 2-quart canteen.

Note. A 2-percent solution of tincture of iodine is a substitute for iodine tablets. Five drops of 2-percent iodine liquid are equivalent to one iodine tablet.

---

   c. Put the cap on the canteen. Shake the canteen to dissolve the tablets.
   d. Wait 5 minutes. Loosen the cap and tip the canteen to allow any leakage to flow away from around the canteen threads.
   e. Tighten the cap and wait an additional 25 minutes before drinking.

15-40. The only approved method of <u>human waste disposal</u> is the <u>saddle trench</u>. The straddle trench latrine is useful when the unit remains in one place for up to 3 days. Each trench is dug 1 foot (30 centimeters) wide, 2-1/2 feet (76 centimeters) deep, and at least 4 feet (1-1/4 meters) long. Multiple trenches are dug at least 2 feet (61 centimeters) apart. Each 4-foot (1.25-meter) trench accommodates two Soldiers. The excavated dirt remains at one end of the latrine for Soldiers to retrieve with shovels in order to cover their excrement and toilet paper. A trench that fills to within 1 foot (30 centimeters) of the top requires the closure of that latrine.

Appendix A

Resources

This appendix provides a quick reference for some of the techniques necessary for Ranger use. It discusses additional information for some methods initially described in chapter content. Finally, it provides an outline on SE that was not part of the main body of this publication.

TRAINING BOARDS

A-1. Figures A-1 through A-10 on pages A-2 through A-11 all depict squad-level tasks. Figures A-11 and A-12 on pages A-12 and A-13 and tables A-1 and A-2 on pages A-14 and A-15 depict information vital for a successful platoon raid. Figures A-13 and A-14 on pages A-16 and A-18 respectively and table A-3 on page A-17 contain information on how to conduct a platoon-level ambush.

A-2. After the figures and tables, this appendix discusses in depth various techniques and formations useful for clearing buildings. It also outlines SE to serve as a handy guide.

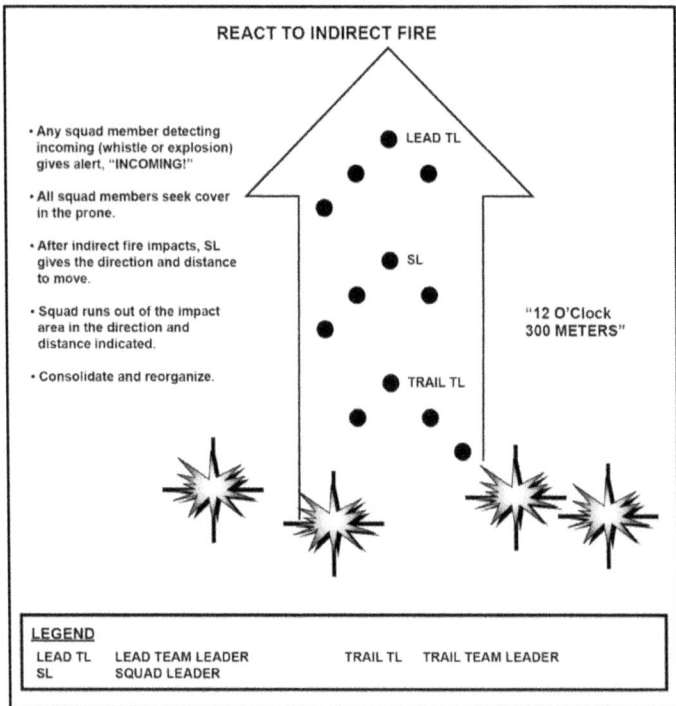

REACT TO INDIRECT FIRE

- Any squad member detecting incoming (whistle or explosion) gives alert, "INCOMING!"

- All squad members seek cover in the prone.

- After indirect fire impacts, SL gives the direction and distance to move.

- Squad runs out of the impact area in the direction and distance indicated.

- Consolidate and reorganize.

LEAD TL

SL

"12 O'Clock 300 METERS"

TRAIL TL

LEGEND

| LEAD TL | LEAD TEAM LEADER | TRAIL TL | TRAIL TEAM LEADER |
| SL | SQUAD LEADER | | |

Figure A-1. React to indirect fire training board

REACT TO DIRECT FIRE CONTACT

- Seek nearest cover.

- Return fire (known or suspected enemy position).

- TM LDRs control fire by using fire commands.

- Report enemy situation (number/position).

- Maintain contact (visual/oral) with team member.

- SQD LDR moves to team in contact and makes an assessment of the situation.

- Factors of his assessment:
  - Can squad move out to engagement area?
  - Can squad gain and maintain suppressive fire?
  - Location of enemy.
  - Size of enemy.
  - Vulnerable flanks.
  - Convered/concealed flanking routes.

- SQD LDR determines COA (break contact or squad attack, for example).

- Report situation to PL.

LEGEND

AR	AUTOMATIC RIFLEMAN	R	RIFLEMAN
COA	COURSE OF ACTION	SQD	SQUAD
GR	GRENADIER	TL	TEAM LEADER
LDR	LEADER	TM	TEAM
PL	PLATOON LEADER		

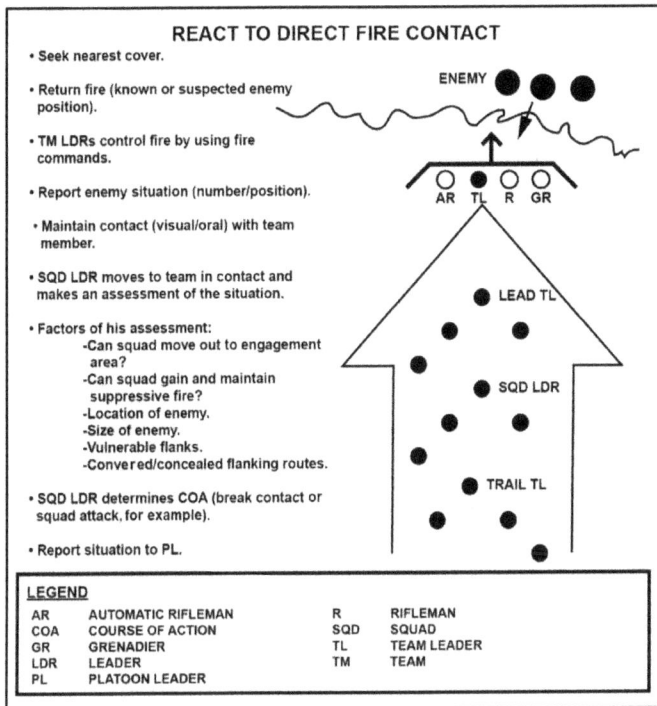

Figure A-2. React to direct fire contact training board

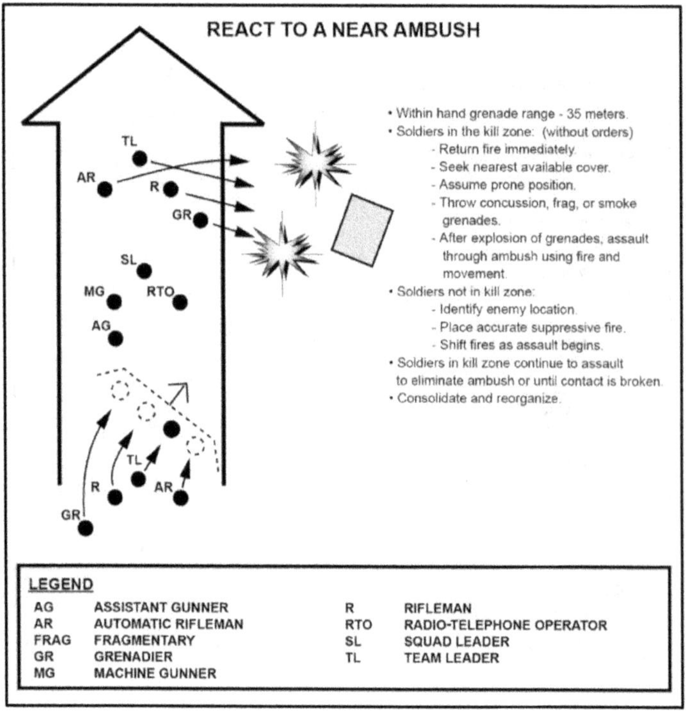
REACT TO A NEAR AMBUSH
TL
AR
R
GR
SL
MG
RTO
AG
TL
R
AR
GR
• Within hand grenade range - 35 meters.
• Soldiers in the kill zone: (without orders)
- Return fire immediately.
- Seek nearest available cover.
- Assume prone position.
- Throw concussion, frag, or smoke grenades.
- After explosion of grenades, assault through ambush using fire and movement.
• Soldiers not in kill zone:
- Identify enemy location.
- Place accurate suppressive fire.
- Shift fires as assault begins.
• Soldiers in kill zone continue to assault to eliminate ambush or until contact is broken.
• Consolidate and reorganize.
LEGEND
AG ASSISTANT GUNNER
AR AUTOMATIC RIFLEMAN
FRAG FRAGMENTARY
GR GRENADIER
MG MACHINE GUNNER
R RIFLEMAN
RTO RADIO-TELEPHONE OPERATOR
SL SQUAD LEADER
TL TEAM LEADER

Figure A-3. React to a near ambush training board

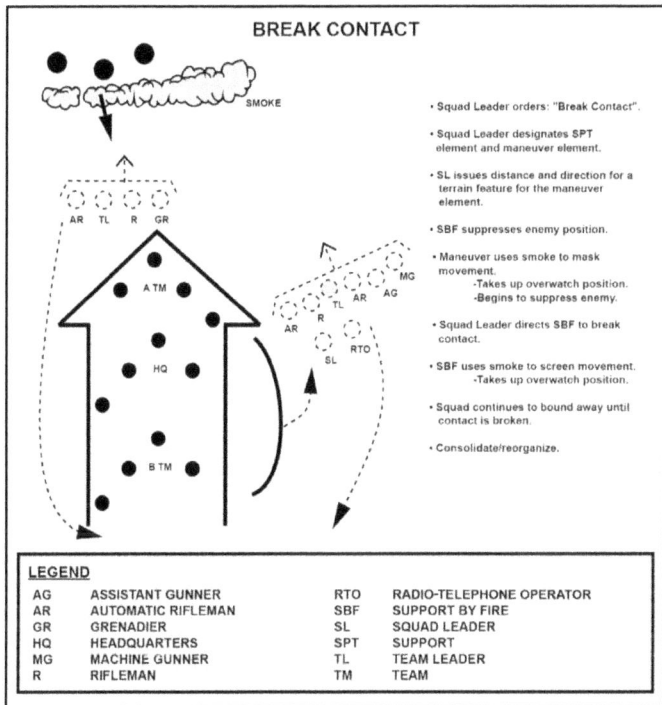

BREAK CONTACT

SMOKE

- Squad Leader orders: "Break Contact".

- Squad Leader designates SPT element and maneuver element.

- SL issues distance and direction for a terrain feature for the maneuver element.

- SBF suppresses enemy position.

- Maneuver uses smoke to mask movement.
  - -Takes up overwatch position.
  - -Begins to suppress enemy.

- Squad Leader directs SBF to break contact.

- SBF uses smoke to screen movement.
  - -Takes up overwatch position.

- Squad continues to bound away until contact is broken.

- Consolidate/reorganize.

LEGEND

AG	ASSISTANT GUNNER	RTO	RADIO-TELEPHONE OPERATOR
AR	AUTOMATIC RIFLEMAN	SBF	SUPPORT BY FIRE
GR	GRENADIER	SL	SQUAD LEADER
HQ	HEADQUARTERS	SPT	SUPPORT
MG	MACHINE GUNNER	TL	TEAM LEADER
R	RIFLEMAN	TM	TEAM

Figure A-4. Break contact training board

FORMATIONS AND ORDER OF MOVEMENT

I. Movement Formation: Fire Team Wedge; MG Team attached.
II. Three Movement Techniques used:
      A. Traveling technique used behind FFL when contact is not likely.
      B. Traveling Overwatch forward of the FFL when enemy contact is possible.
      C. Bounding Overwatch used forward of the FFL when enemy contact is expected.
III. Distances are based on but not dictated by visibility, terrain, and vegetation.
IV. Actions at Night: Modified Wedge
V. Actions at the Halt: Short and Long Halt (GV/LV)
VI. Leader Location: Fixed/Unfixed

FIRE TEAM WEDGE

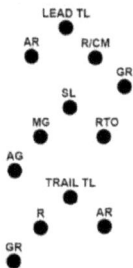

LEAD TL

AR   R/CM

GR

SL

MG   RTO

AG

TRAIL TL

R   AR

GR

MODIFIED WEDGE

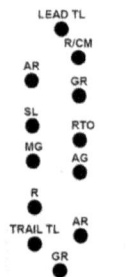

LEAD TL

R/CM

AR   GR

SL

MG   RTO

AG

R

TRAIL TL   AR

GR

LEGEND

AG	ASSISTANT GUNNER	R	RIFLEMAN
AR	AUTOMATIC RIFLEMAN	R/CM	RIFLEMAN/COMPASS MAN
FFL	FORWARD FRIENDLY LINE	RTO	RADIO-TELEPHONE OPERATOR
GR	GRENADIER	SL	SQUAD LEADER
GV/LV	GOOD/LIMITED VISIBILITY	TL	TEAM LEADER
MG	MACHINE GUNNER		

Figure A-5. Formations and order of movement training board

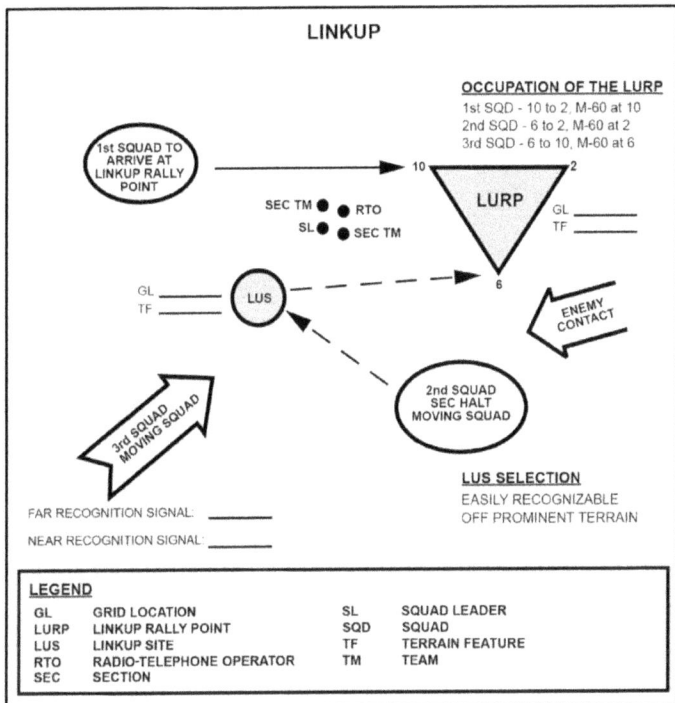

Figure A-6. Linkup training board

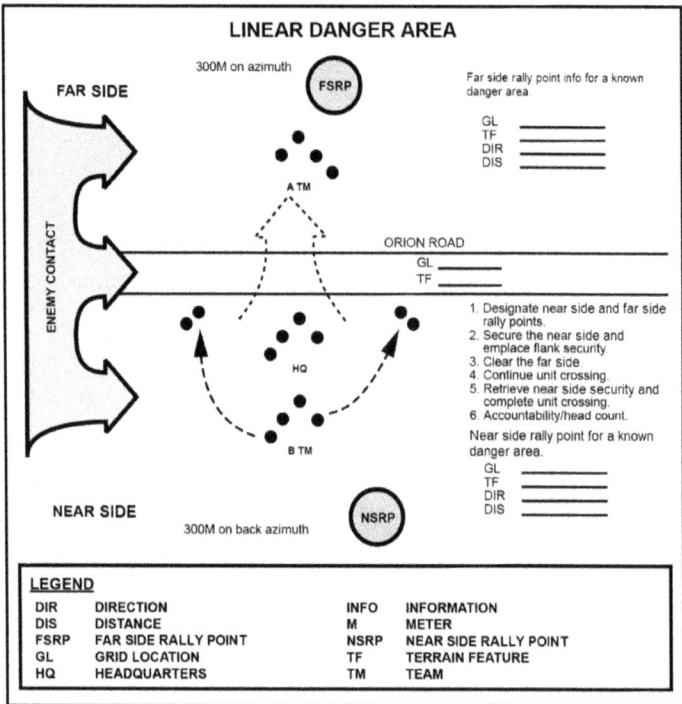

Figure A-7. Linear danger area training board

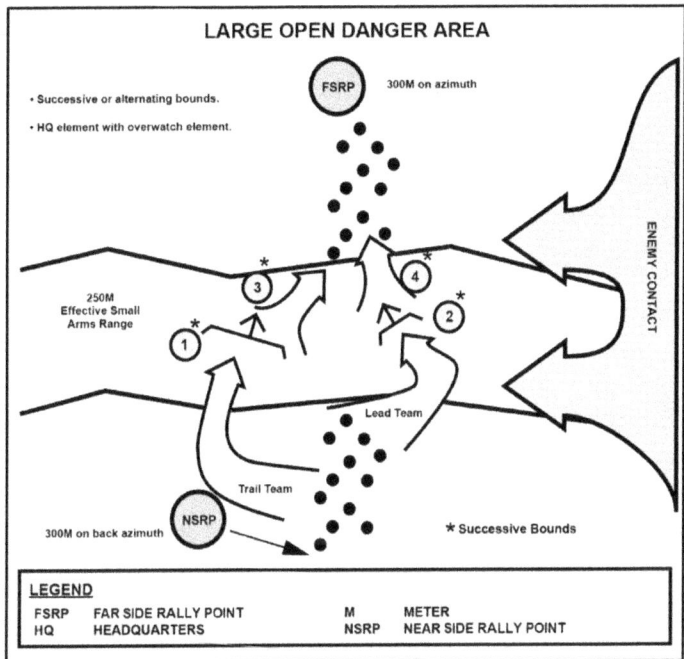

LARGE OPEN DANGER AREA

- Successive or alternating bounds.
- HQ element with overwatch element.

FSRP — 300M on azimuth

250M Effective Small Arms Range

ENEMY CONTACT

Lead Team

Trail Team

NSRP

300M on back azimuth

★ Successive Bounds

LEGEND

FSRP	FAR SIDE RALLY POINT		M	METER
HQ	HEADQUARTERS		NSRP	NEAR SIDE RALLY POINT

Figure A-8. Large, open danger area training board

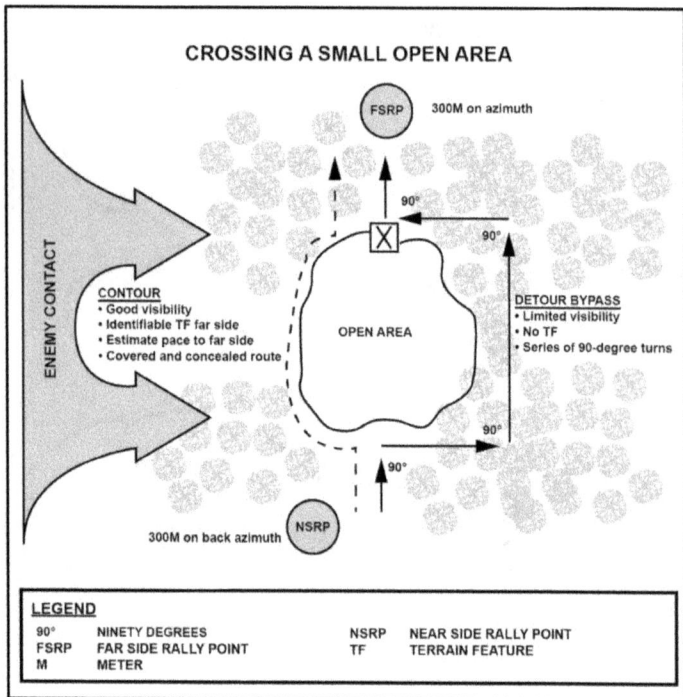

Figure A-9. Crossing a small, open area training board

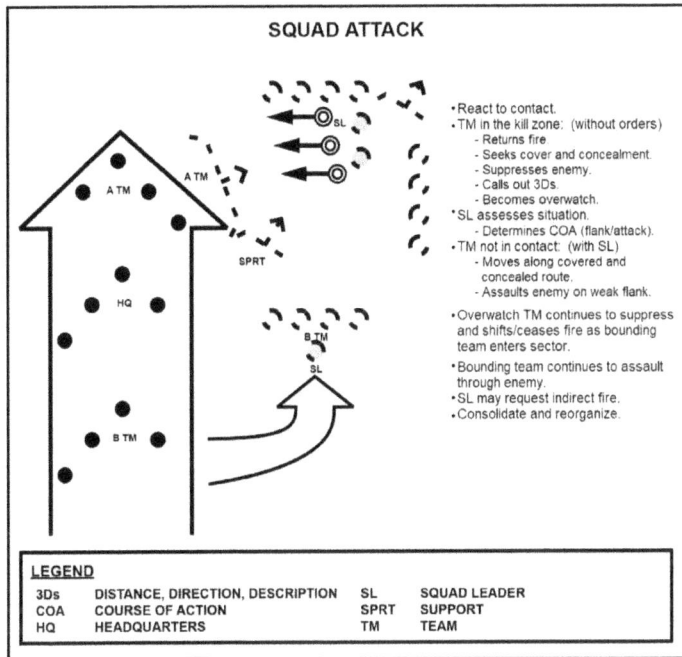

Figure A-10. Squad attack training board

RAID BOARDS (LEFT)
GENERAL INFO

Raid References:
- RHB Chapter 7
- ATP 3-21.8

PURPOSES:
1. Destroy
2. Liberate
3. Collect Intelligence

Planning Considerations:
- Mission
- Enemy
- Troops
- Terrain/OAKOC
- Time
- Civilians

STUDENT INPUT WRITTEN HERE

Raid:
A surprise attack on a fixed position or installation ending in a planned withdrawal.

Characteristics:
1. Surprise - Gain
2. Coordinated fires - Maintain
3. Violence of actions - Retain
4. Planned withdrawal

INITIATIVE

LEGEND

ATP   ARMY TRAINING PUBLICATION   RHB   RANGER HANDBOOK

OAKOC - observation and fields of fire, avenues of approach, key terrain, obstacles, and cover and concealment

Figure A-11. Raid boards (left)

RAID BOARDS (MIDDLE)
ACTIONS ON OBJECTIVE

Prep Clearing Team:
PL, RTO, WSL,
3 ABs, SEC TM

ORP:
Leader's Recon
Prep MWE

SECURITY HALT

Leader's Recon
•Pinpoint OBJ
•Finalize ORP
•Determine Enemy Situation
•Emplace Surveillance
•Emplace Security
•Reconnoiter Positions and Routes
•ID Control Measures
•ID and Prioritize Targets
•Left and Right Limits, Assault and Support
•Divide OBJ
•Signals
•LOA
•Plan to Secure OBJ
•Confirm Plan

RP

ASSLT POS

LSS

RSS

FP

FP

TENT

LEGEND

AB	AMMUNITION BEARER	ORP	OBJECTIVE RALLY POINT
ASSLT POS	ASSAULT POSITION	PL	PLATOON LEADER
FP	FIRING POINT	RECON	RECONNAISSANCE
ID	IDENTIFY	RP	RALLY POINT
LOA	LIMIT OF ADVANCE	RSS	RIGHT SIDE SECURITY
LSS	LEFT SIDE SECURITY	RTO	RADIO-TELEPHONE OPERATOR
MWE	MEN, WEAPONS, EQUIPMENT	SEC TM	SECURITY TEAM
OBJ	OBJECTIVE	WSL	WEAPONS SQUAD LEADER

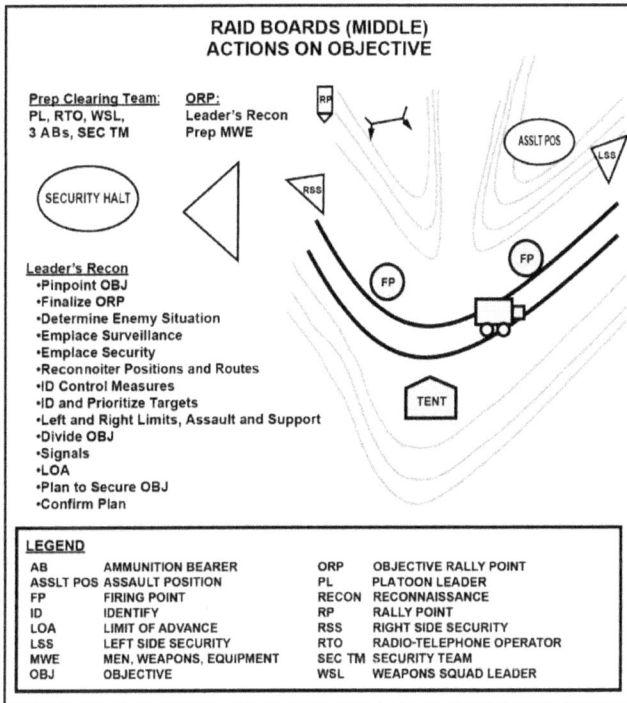

Figure A-12. Raid boards (middle)

Table A-1. Raid boards (right) task organization

ELEMENT	WHO	WHY	WHAT
HQ	PL, PSG, RTO, FO, Medic	C2, Mission command	One AN/PRC-119F radio Two AN/PRC-148 radios One AN/PSN-13 navigation set One AN/PVS-14 NVD (per individual) One AN/PEQ-15 ATPIAL (per weapon)
ASSLT 1	Squad (+) (–)	Destroy	Three AN/PRC-148 radios One AN/PSN-13 navigation set Two M18A1 claymore mines One AT4 (antitank) weapon One AN/PVS-14 NVD (per individual) One AN/PEQ-15 ATPIAL (per weapon)
ASSLT 2	Squad (+) (–)	Destroy	Three AN/PRC-148 radios One AN/PSN-13 navigation set Two M18A1 claymore mines One AT4 (antitank) weapon One AN/PVS-14 NVD (per individual) One AN/PEQ-15 ATPIAL (per weapon)
Support	Weapons Squad (+) (–)	Suppress	Three M240B machine guns Three AN/PRC-148 radios One AN/PSN-13 navigation set One AN/PVS-14 NVD (per individual) One AN/PEQ-15 ATPIAL (per weapon)
Security	Squad (+) (–)	Delay	Two AN/PRC-148 radios One AN/PSN-13 navigation set One AN/PVS-14 NVD (per individual) One AN/PEQ-15 ATPIAL (per weapon)

Legend: ASSLT–assault; ATPIAL–Advanced Target Pointer / Illuminator / Aiming Laser; C2–command and control; FO–forward observer; HQ–headquarters; NVD–night-vision device; PL – platoon leader; PSG–platoon sergeant; RTO–radio-telephone operator

Table A-2. Raid boards (right) standard operating procedure

RAID BOARDS (RIGHT) SOP	
CONTINGENCY	**SOP**

1. Contingency
  a. Objective rally point
  b. Leader's reconnaissance
  c. Occupation
2. Mass casualties
3. Counterattack
	1. Enemy prisoner of war search
	2. Aid and litter team
	3. Medical evacuation
	4. Casualty collection point
	5. Withdrawal plan

Legend: SOP–standard operating procedure	

AMBUSH SOP (LEFT)

References:
• RHB Chapter 7
• ATP 3-21.8

Ambush:
A surprise attack from a covered and concealed position on a moving or temporarily halted target.

Types:
• Point
• Area

Fundamentals:
1. Surprise - Gain
2. Coordinated fires - Maintain
3. Violence of actions - Retain

PURPOSES:
1. Disrupt/Destroy
2. Collect Intelligence
3. Block or Deny Access
4. Canalize

Raid:
Deliberate: Platoon has specific target at a predetermined time and location.

Hasty: Platoon makes visual contact with the enemy and has time to establish an ambush without being detected.

Planning Considerations:
• Mission - Task/Purpose
• Enemy - MPCOA, MDCOA
• Terrain - Map Recon/Ldrs Recon
• Time - 1/3 - 2/3 Rule
• Troops - Task Org
• Civilians

LEGEND

ATP	ARMY TRAINING PUBLICATION	ORG	ORGANIZATION
LDRS	LEADERS	RECON	RECONNAISSANCE
MDCOA	MOST DANGEROUS COURSE OF ACTION	RHB	RANGER HANDBOOK
MPCOA	MOST PROBABLE COURSE OF ACTION	SOP	STANDARD OPERATING PROCEDURE

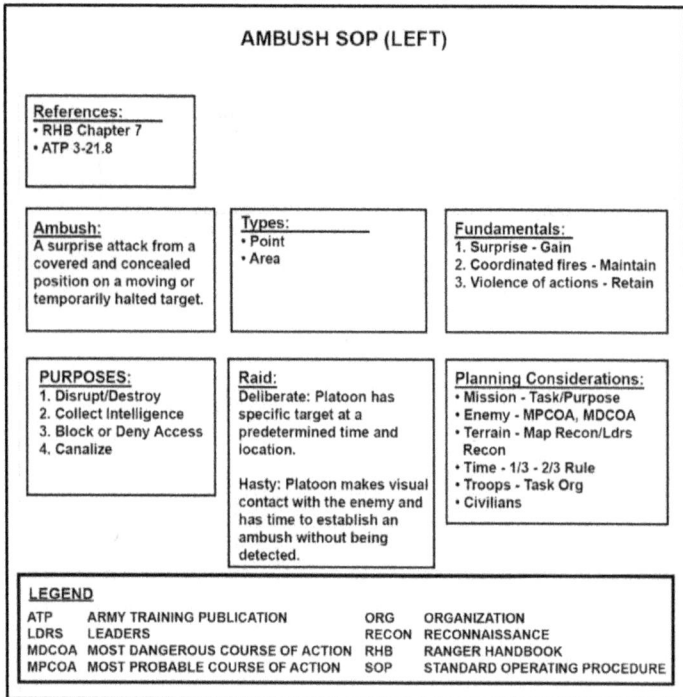

Figure A-13. Ambush standard operating procedure (left)

Table A-3. Ambush boards (middle)

ELEMENT	WHO	WHY	WHAT
HQ	PL, PSG, RTO, FO, Medic	C2, Mission Command	One AN/PRC-119F radio Two AN/PRC-148 radios One AN/PSN-13 navigation set One AN/PVS-14 NVD (per individual) One AN/PEQ-15 ATPIAL (per weapon)
Security	Minimum of three Two-person teams	Early Warning Seals off OBJ LSS, RSS, ORP Security	Three AN/PRC-148 radios One AN/PSN-13 navigation set Two M18A1 claymore mines One AT4 (antitank) weapon One AN/PVS-14 NVD (per individual) One AN/PEQ-15 ATPIAL (per weapon)
Support	Weapons squad, WSL	Suppressive Fire Security on OBJ during RECON/REORG	Three M240B machine guns Three AN/PRC-148 radios One AN/PSN-13 navigation set One AN/PVS-14 NVD (per individual) One AN/PEQ-15 ATPIAL (per weapon)
Assault	Two squads (+) (−)	Main Effort Block, Destroy, Canalize, Capture PIR	Three AN/PRC-148 radios One AN/PSN-13 navigation set Two M18A1 claymore mines One AT4 (antitank) weapon One AN/PVS-14 NVD (per individual) One AN/PEQ-15 ATPIAL (per weapon)

Legend: ATPIAL–Advanced Target Pointer/Illuminator/Aiming Laser; C2–command and control; FO–forward observer; HQ–headquarters; LSS–left side security; NVD–night-vision device; OBJ–objective; ORP–objective rally point; PIR–priority intelligence requirement; PL–platoon leader; PSG–platoon sergeant; RECON/REORG–reconnaissance/reorganization; RSS–right side security; RTO–radio-telephone operator; WSL–weapons squad leader

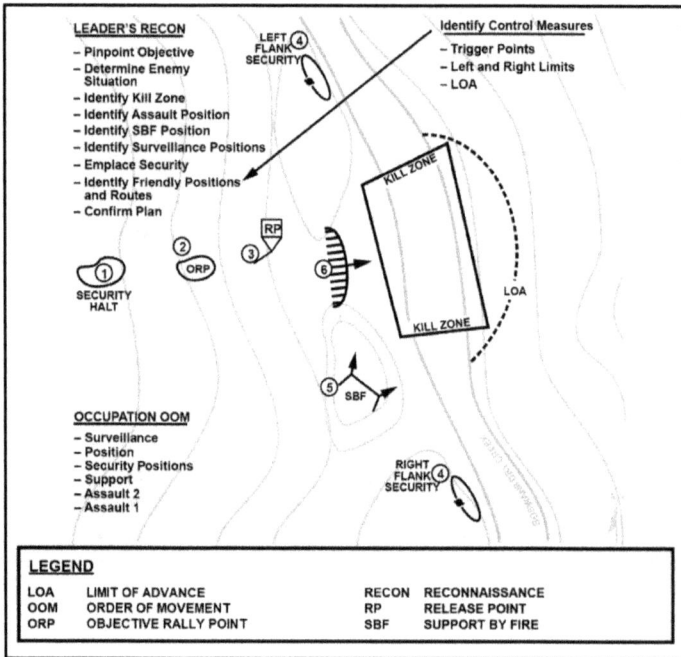

Figure A-14. Leader's reconnaissance training board

TC 3-21.76

BUILDING-CLEARING TECHNIQUES

A-3. Rangers need to understand the fundamentals of urban operations, part of which is moving as a unit through a building. Urban buildings have much larger architectural dimensions and much more complicated floor plans and interior designs. Safely searching and clearing their interiors requires a thorough, organized effort that minimizes the chances of surprise and fratricide. Paragraphs A-4 through A-10 cover techniques for maneuvering through just such an urban environment and techniques for clearing various types of rooms.

CLEARING ROOMS

A-4. On the signal, the team enters through the entry point or breach. As the team members move to their points of domination, they engage all threats or hostile targets in sequence in their sectors. The direction each Ranger moves is not preplanned unless the exact room layout is known. However, each Ranger moves in the opposite direction of the preceding Ranger. The following procedure provides a demonstrative example of this:

- Ranger #1 enters the room and eliminates any immediate threat. Then, Ranger #1 moves left or right along the path of least resistance to a point of domination—one of the two corners—and continues down the room to gain depth.
- Ranger #2 enters almost simultaneously with Ranger #1 but moves in the opposite direction, following the wall. Ranger #2 clears the entry point, clears the immediate threat area, and moves to a point of domination.
- Ranger #3 simply moves in the opposite direction of Ranger #2 inside the room, moves at least 1 meter (3 feet) from the entry point, and takes a position that dominates the sector.
- Ranger #4 moves in the opposite direction of Ranger #3, clears the doorway by at least 1 meter (3 feet), and moves to a position that dominates the sector. (See figure 8-9 on page 8-19, figure 8-10 on page 8-20, and figure 8-11 on page 8-22.)

DIAMOND FORMATION (SERPENTINE TECHNIQUE)

A-5. The serpentine technique is a variation of a diamond formation that is useful in a narrow hallway. Ranger #1 provides security to the front. The sector of fire includes any enemy soldiers appearing at the far end or along the hallway. Rangers #2 and #3 cover the left and right sides of Ranger #1. Their sectors of fire include any enemy combatants appearing suddenly from either side of the hall. Ranger #4, who normally carries the M249 machine gun, provides rear protection against any enemy soldiers appearing behind the team. (See figure A-15 on page A-20.)

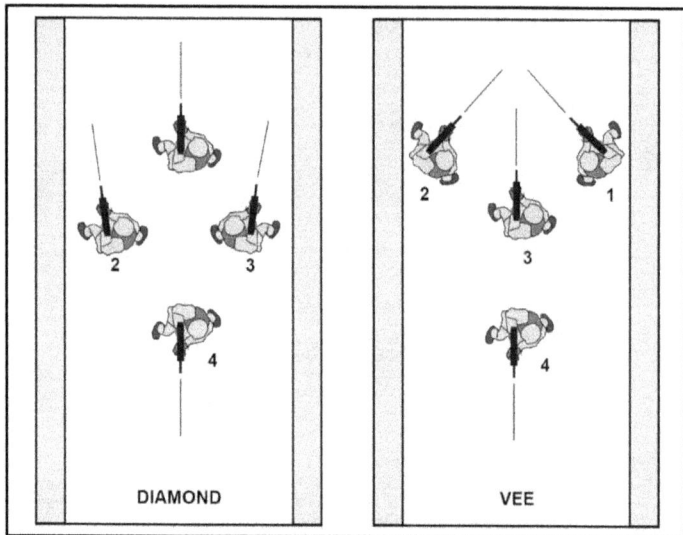

Figure A-15. Diamond and vee formations

VEE FORMATION (ROLLING-T TECHNIQUE)

A-6. The rolling-T technique is a variation of the vee formation and is useful in wide hallways. Rangers #1 and #2 move abreast, covering the opposite side of the hallway from the one they are walking. Ranger #3 covers the far end of the hallway from a position behind Rangers #1 and #2, firing between them. Ranger #4 provides rear security. (See figure A-15.)

Resources

CLEARING HALLWAY JUNCTIONS

A-7. Hallway intersections are danger areas requiring a cautious approach. Figure A-16 on page A-22 depicts the fire team's actions upon reaching a T-intersection when approaching along the cross of the T. The unit is using the diamond (serpentine) formation for movement. (See figure A-16 on page A-22, step A.) To clear a hallway—

- The team configures into a modified 2-by-2 (box) formation with Rangers #1 and #3 abreast and toward the right side of the hall. Ranger #2 moves to the left side of the hall and orients to the front, and Ranger #4 shifts to the right side of the hall (left arm toward the wall) and maintains rear security. When clearing a right-hand corner, use the left-handed firing method to minimize exposure. (See figure A-16 on page A-22, step B.)
- Rangers #1 and #3 move to the edge of the corner. Ranger #3 assumes a low crouch or kneeling position. On signal, Ranger #3, keeping low, turns right around the corner, and Ranger #1, staying high, steps forward while turning to the right. (The sectors of fire interlock, and the combination of low and high positions prevents Soldiers from firing at one another.) (See figure A-16 on page A-22, step C.)
- Rangers #2 and #4 continue to move in the direction of travel. As Ranger #2 passes behind Ranger #1, Ranger #1 shifts laterally to the left until reaching the far corner. (See figure A-16 on page A-22, step D.)
- Rangers #2 and #4 continue to move in the direction of travel. As Ranger #4 passes behind Ranger #3, Ranger #3 shifts laterally to the left until reaching the far corner. As Ranger #3 begins to shift across the hall, Ranger #1 turns into the direction of travel and moves to the original position in the diamond (serpentine) formation. (See figure A-16 on page A-22, step E.)
- As Rangers #3 and #4 reach the far side of the hallway, they, too, assume their original positions in the diamond (serpentine) formation, and the fire team continues to move. (See figure A-16 on page A-22, step F.)

Figure A-16. Movement formations to clear hallway junctions

CLEARING T-INTERSECTIONS

A-8. Figure A-17 on page A-24 depicts the fire team's actions upon reaching a T-intersection when approaching from the base of the T. The fire team is using the diamond (serpentine) formation for movement. (See figure A-17 on page A-24, step A.) To clear a T-intersection—

- The team configures into a 2-by-2 (box) formation with Rangers #1 and #2 aligning for the left side and Rangers #3 and #4 aligning for the right side. When clearing a right-hand corner, use the left-handed firing method to minimize exposure. (See figure A-17 on page A-24, step B.)
- Rangers #1 and #3 move to the edge of the corner and assume a low crouch or kneeling position. On signal, Rangers #1 and #3 simultaneously turn left and right, respectively. (See figure A-17 on page A-24, step C.)
- At the same time, Rangers #2 and #4 step forward and turn left and right, respectively, while maintaining their (high) positions. (The sectors of fire interlock, and the combination of low and high positions prevents Soldiers from firing at one another.) (See figure A-17 on page A-24, step D.)
- Once the left and right portions of the hallway are clear, the fire team resumes the movement formation. (See figure A-17 on page A-24, step E.) Unless the team leaves behind security, the hallway will not remain clear once the fire team leaves the immediate area.

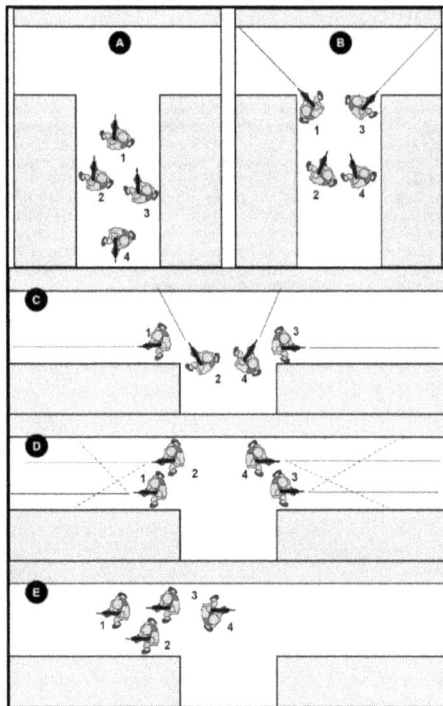

Figure A-17. Movement formations to clear T-intersections

CLEARING STAIRWELLS AND STAIRCASES

A-9. Stairwells and staircases are comparable to doorways because they create a fatal funnel. The three-dimensional aspect of additional landings intensifies the danger. The ability of units to conduct the movement depends upon the direction they are traveling and the layout of the stairs. Regardless, the clearing technique follows a basic format:

- The leader designates an assault element to clear the stairs.
- The unit maintains 360-degree, three-dimensional security in the vicinity of the stairs.
- The leader then directs the assault element to locate, mark, and bypass or clear (or both) any obstacles or booby traps blocking access to the stairs.
- The assault element moves up (or down) the stairway by using either the two-, three-, or four-person flow technique, providing overwatch up and down the stairs while moving. The three-person variation (see figure A-18 on page A-26) is preferable.

A-10. Figure A-18 on page A-26 best portrays the three-person flow clearing technique. All Rangers work together, covering each other and focusing on their sectors of fire.

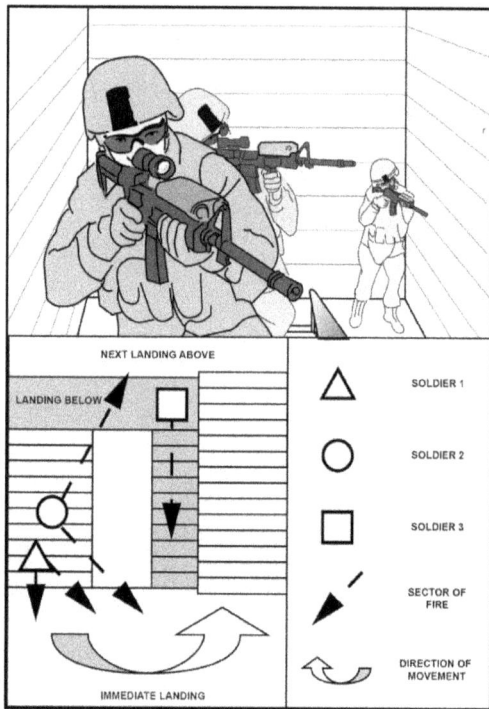

Figure A-18. Three-person flow clearing technique

FOLLOWTHROUGH

A-11. After securing a floor (bottom, middle, or top), the leader selects and assigns unit members to cover potential enemy counterattack routes to the building. The leader gives priority initially to securing the direction of attack. Security elements alert the unit and place a heavy volume of fire on enemy forces approaching the unit. Units guard all AAs leading into their areas. These include—

- Enemy mouseholes between adjacent buildings.
- Covered routes to the building.
- Underground routes into the basement.
- Approaches over adjoining roofs or from window to window.

A-12. Units who performed missions as assault elements remain prepared to assume an overwatch mission and to support another assault element. To continue the mission—

- Units maintain momentum. This is a critical factor in clearing operations. Units do not allow the enemy to move to their next set of prepared positions or to prepare new positions.
- The support element pushes replacements, ammunition, and supplies forward to the assault element.
- Units evacuate and replace their casualties.
- Units establish security for cleared areas in accordance with the OPORD or tactical SOP.
- Units mark all cleared areas and rooms in accordance with the unit's SOP.
- The support element displaces forward to ensure it is in place to provide support to the assault element, such as in the isolation of the new objective.

SITE EXPLOITATION

A-13. Sensitive SE is the collection and analysis of information, materials, and persons from a designated location to answer information requirements, facilitate subsequent operations, or support criminal prosecution. The duties and responsibilities of the SE members follow:

- TL:
  - Gathers mission planning products.
  - Briefs sensitive SE during the WARNORD and OPORD.
  - Preplans objective search priorities.
  - Establishes premission checks and rehearsals.
  - Coordinates exploitation tasking and prioritization.
  - Establishes priority rooms.
  - Directs activities on the target.
  - Orchestrates and leads the backbrief.
- Assistant TL:
  - Conducts an inventory of the sensitive SE kit.
  - Conducts precombat inspections.
  - Conducts rehearsals and rock drills with the TL.
  - Determines the marshalling area.
  - Determines the evidence placement point.
  - Prepares and packages evidence for transfer.
  - Sketches the objective.
  - Records events.

- Searchers (working in two-person teams as feasible):
  - Train any additional searchers.
  - Assist with sensitive SE kit inventories.
  - Conduct precombat inspections.
  - Conduct rehearsals.
  - Assist with the packaging and labeling of evidence.
  - Communicate all findings to the TL, tactical questioner, or both.
- Marshalling officer:
  - Conducts precombat inspections on equipment.
  - Conducts a detailed search of the marshalling area.
  - Collects biometrics on those in the marshalling area.
- Tactical questioner:
  - Assists with the collection of detainees' biometrics.
  - Communicates the collected PIR to the TL.
  - Helps searchers with information gathering and organization.
  - Prioritizes all personnel under control.
  - Prepares questions during precombat inspections.

A-14. During the initial assessment, the team conducts 5- and 25-meter (16- to 82-foot) checks around the target, looking toward and away from the target building. The TL determines how much time is left on target, what support is available, who is on target, and what has been found on target, while the assistant TL determines the marshalling area. The TL and assistant TL conduct the initial walkthrough of the target; the TL has the assistant TL walk through the target building and discuss what was found on target. The TL labels rooms while moving through the target; the assistant TL labels rooms when the TL is busy. The assistant TL sketches the target to include the priority of the rooms based on the walkthrough and PIRs. The TL prioritizes the rooms in the order of the importance of their search.

A-15. During a room and building search, the TL determines the priority area. The team searches this area first and then completes a systematic, 360-degree search of the entire room. Conduct in the searching of a room is as follows:

- Use clockwise movement and a low, then medium, and finally high search pattern to cover all areas. Label any drop holes and false walls.
- Move all cleared items such as tables, chairs, and pots to the center of the room.
- Place evidence bags in the evidence placement point for use while searching the room. Later, move items from the individual room's evidence placement point to the centralized evidence placement point.
- At least two people search each room as follows:
  - Mark the label of the room with the letter $\underline{X}$ upon the completion of its search. The X-mark confirms two people have searched the room.
  - Search all remaining rooms in clockwise order at their low, medium, and high points.
- The assistant TL continues to sketch, identifying dead space in the building's evidence placement point and ensuring the point is in an already searched and cleared area.

A-16. The marshalling officer and TQ person clear the marshalling area, set up two pits, and then clear the area where the unit found the detainees on target. The two pits are called dirty and clean. The dirty pit is for those detainees searched once for weapons. The clean pit is for those detainees who have undergone a detailed search for any items.

A-17. In the marshalling area, the marshalling officer and TQ person search all the detainees, collect the detainees' biometrics, separate the detainees, and relay all information to the TL. They conduct two searches. The first search ensures the detainees do not have weapons and occurs in the dirty pit. At this time, the team places flex cuffs and a detainee bag on the detainee. The second search is a systematic search of a detainee's person—from top to bottom, front to back, and left to right—in the clean pit, a search that also examines every article of clothing. When collecting detainees' biometrics, the marshalling officer and TQ person collect fingerprints, an iris scan, and a photo with the secure biometric live scan device or a ten-print card. They collect all media devices for further exploitation. The TQ person separates individuals from the other detainees to question further. The TQ person and marshalling officer relay all pertinent information to the TL to assist in decision making.

A-18. Upon mission completion, the TL collapses the objective and conducts a final walkthrough to ensure no signs of U.S. tactics, techniques, and procedures remain on target. The TL ensures accountability of personnel, weapons, equipment items, and detainees found on target and prepares for exfiltration.

This page intentionally left blank.

Appendix B
Quick Reference Cards

The intent of this appendix is for quick reference for Rangers in training or the field. It provides illustrations or tabular information useful in completing various forms and formatted reports frequently required of Rangers.

REFERENCE CARDS

B-1. Figures B-1 through B-4, appearing across pages B-2 through B-5, depict critical reference cards Rangers routinely use during missions. These include a reference card for IEDs and unexploded explosive ordnances, DA Form 5517 (*Standard Range Card*), sample formats of the 9-line MEDEVAC request and the MIST report. Rangers should also familiarize themselves with DD Form 1380. Table B-1 on pages B-6 through B-8 explains in greater depth the content for the line items in the 9-line MEDEVAC request format that chapter 15 initially introduced.

B-2. Although this publication introduced some of them earlier, these reference cards in this appendix serve as handy reminders for Rangers in the field. Check periodically for updated information on the Army Training Network.

IED / UXO

Procedures when IEDs are found

Security - Maintain 360 degrees. Scan close in and far out, up high and down low.

Always - Scan your immediate surroundings for more IEDs.

Move - Move away. Vary safe distances, but plan for 300m minimum safe distance and adapt to your METT-TC (I).

Attempt - To confirm suspected IEDs using optics while staying back as far as possible.

Cordon - Off the area. Direct people out of danger area. Do not allow anyone to enter except for EOD. Question, search, and detain suspects as defined by your existing ROE.

Report - Your situation using the 9-line IED / UXO spot report format.

This could be your hand if you try to dispose of UXOs or IEDs. The enemy has developed anti-handling to catch you when you try defusing. Leave it to experts!

Call EOD - Don't be a Hero!

IED / UXO Report

LINE 1. DATE-TIME-GROUP: when the item was discovered.

LINE 2. REPORT ACTIVITY AND LOCATION: Unit and grid location of the IED/UXO.

LINE 3. CONTACT METHOD: Radio frequency, call sign, POC, and telephone number.

LINE 4. TYPE OF ORDNANCE: Dropped, projected, placed, or thrown. Give the number of items if more than one.

LINE 5. NBC CONTAMINATIONS: Be as specific as possible.

LINE 6. RESOURCES THREATENED: Equipment, facilities, or other assets that are threatened.

LINE 7. IMPACT ON MISSION: Short description of current tactical situation and how the IED/UXO affects the status of the mission.

LINE 8. PROTECTIVE MEASURES: Any measures taken to protect personnel and equipment.

LINE 9. RECOMMENDED PRIORITY: Immediate, Indirect, Minor, No Threat.

Priority

Immediate: Stops unit's maneuver and mission capability or threatens critical assets vital to the mission.

Indirect: Stops the unit's maneuver and mission capability or threatens critical assets important to the mission.

Minor: Reduces the unit's maneuver and mission capability or threatens critical assets of value.

No Threat: Has little or no effect on the unit's capabilities or assets.

LEGEND

EOD	EXPLOSIVE ORDNANCE DISPOSAL	NBC	NUCLEAR, BIOLOGICAL, AND CHEMICAL
IED	IMPROVISED EXPLOSIVE DEVICE	POC	POINT OF CONTACT
m	METER	ROE	RULES OF ENGAGEMENT
METT-TC (I)	MISSION, ENEMY, TERRAIN / WEATHER, TROOPS / SUPPORT / TIME, CIVIL / INFORMATIONAL CONSIDERATIONS	UXO	UNEXPLODED EXPLOSIVE ORDNANCE

Figure B-1. Improvised explosive device or unexploded explosive ordnance reference card

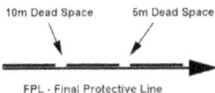

Figure B-2. Example of DA Form 5517, Standard Range Card (with final protective line)

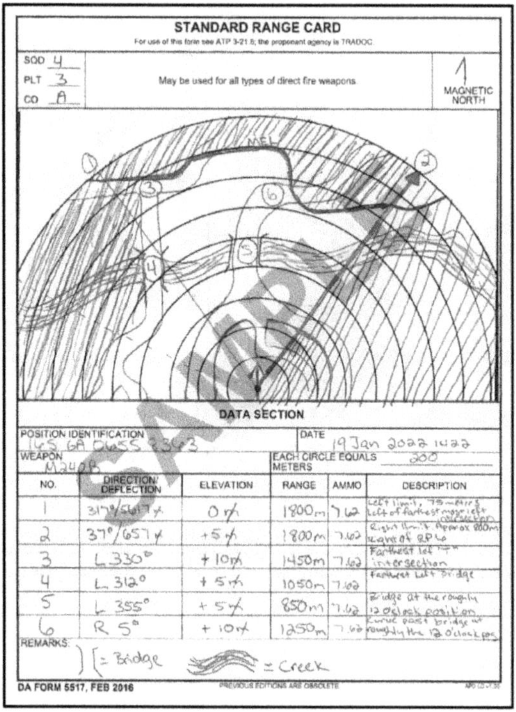

STANDARD RANGE CARD

For use of this form see ATP 3-21.8; the proponent agency is TRADOC.

SQD 4
PLT 3
CO A

May be used for all types of direct fire weapons.

MAGNETIC NORTH

DATA SECTION

POSITION IDENTIFICATION					DATE
U.S GA Class 230-3					19 Jan 2022 1422
WEAPON				EACH CIRCLE EQUALS	
M240B				300	METERS

NO.	DIRECTION/ DEFLECTION	ELEVATION	RANGE	AMMO	DESCRIPTION
1	31°/5617 ↓	0 ↑	1800m	7.62	Left limit, 78-series left of farthest mover left of terrain
2	37°/657 ↓	+5 ↑	1800m	7.62	Right limit, approx 100m right of RP ↓
3	L 330°	+10 ↑	1450m	7.62	Farthest left intersection
4	L 312°	+5 ↑	1050m	7.62	Farthest left Bridge
5	L 355°	+5 ↑	850m	7.62	Bridge at the roughly 12 o'clock position
6	R 5°	+10 ↑	1250m	7.62	Curve past bridge at roughly the 12 o'clock pos

REMARKS:] = Bridge   ≈ Creek

DA FORM 5517, FEB 2016   PREVIOUS EDITIONS ARE OBSOLETE   APD LC v3.0ES

Figure B-3. Example of DA Form 5517, Standard Range Card

NATO 9-Line MEDEVAC Request Format
(use brevity codes for non-secure communications or use full description for more clarity)

1. Location of Pick-up Site	AB 427349+2			
2. Call Sign & Frequency of Requesting Unit	37500	RANGER 6		
3. # Patients by Precedence				
Urgent (1 hr):	Priority (4 hrs): X	Routine (24 hrs):		
4. Special Equipment Required				
A. None X	B. Hoist	C. Extrication Equip	D. Ventilator	E. Other (describe)
5. # Patients by Type				
I. Litter: 1	A. Ambulatory	E. Escort (women/children/HVT)		
6. Security at Pick-up Site				
N. No Enemy	N			
P. Possible enemy troops in area				
E. Enemy troops in area (approach with caution)				
X. Enemy troops in area (armed escort required)				
7. Method of Marking Pick-up Site				
A. Panel X	B. Pyrotechnic	C. Smoke X	D. None	E. Other (describe)
8. Patient Nationality & Status (# by type)				
A. US/Coalition Military, Nationality:	X			
B. US/Coalition Civilian, Nationality:				
C. Non-US/Coalition Military, Nationality:				
D. Non-US/Coalition Civilian, Nationality:				
E. Enemy Prisoner of War:				
F. High Value Target (escort required):				
9. Terrain Description FLAT, NEAR LAKE				

MIST Report
(Required for Each Patient; Reference Patient's DD 1380 TCCC Card)

Patient ID (i.e. Battle Roster):	525794			
M - Mechanism of Injury i.e. blast, gunshot wound (GSW), etc.; can be NONE if medical complaint	GSW			
I - Injuries Sustained i.e. penetrating wound, laceration, burn, amputation, etc.; include body location	PENETRATING WOUND			
S - Signs and Symptoms				
Pulse	68 R	69 R		
Blood Pressure	120/80	118/80		
Respiratory Rate	20	20		
Level of Consciousness (AVPU)	A-ALERT	A-ALERT		
Other				
T - Treatment Given i.e. tourniquet, NPA, needle D, fluids, medications	FENTANYL 800 MG BTPC 1100	MORPHINE 20 MG IV'S 1115		

Figure B-4. Examples of a 9-line medical evacuation request and a mechanism of injury, injury type, signs, treatment report

Table B-1. Explanations of the nine lines on a medical evacuation request

Line	Item	Explanation	Where and How Obtained	Normal Provider	Reason
1	Location of pickup site	Grid coordinates of the pickup site are sent by secure communication. To prevent confusion, the grid zone letters are included in the message.	From map or navigational device, determine the military grid reference system six- or eight-digit grid coordinates of the pickup site.	Unit leader(s)	Required so the evacuation vehicle knows where to pick up the patient(s). Also required when the evacuation vehicle is picking up from more than one location and the unit coordinating the evacuation mission is planning the route.
2	Radio frequency, call sign, and suffix	Frequency of the radio at the pickup site and not a relay frequency. The call sign and any suffix of the contact person at the pickup site is transmittable in the clear.	From automated net control device or other approved means.	Radio transmission operator	Required to enable the evacuation vehicle to contact the requesting unit while en route and obtain additional information or changes in the situation or directions.
3	Number of patients by precedence	A—URGENT B—URGENT-SURG C—PRIORITY D—ROUTINE E—CONVENIENCE If the same request is reporting two or more categories, insert the word *break* between each category.	From evaluation of patient(s).	Medic or senior person present	Required by the unit controlling the vehicles to assist in prioritizing missions.
4	Special equipment required	A—None B—Hoist C—Extraction equipment D—Ventilator	From evaluation of patient(s) or situation.	Medic or senior person present	Required so the equipment is placed aboard the evacuation vehicle prior to the start of the mission.

Table B-1. Explanations of the nine lines on a medical evacuation request (continued)

Line	Item	Explanation	Where and How Obtained	Normal Provider	Reason
5	Number of patients by type	Report only applicable information. When requesting evacuation for both types, insert the word *break* between the litter and ambulatory entries. L+# of patients–Litter A+# of patients– Ambulatory (sitting)	From evaluation of patient(s).	Medic or senior person present	Required so the appropriate number of evacuation vehicles dispatch to the pickup site and are appropriately configured to carry the patients requiring evacuation.
6	[in wartime] Security at pickup site	N—No enemy troops in area P—Possibly enemy troops in area; approach with caution E—Enemy troops in area; approach with caution X—Enemy troops in area; armed escort required	From evaluation of situation.	Unit leader(s)	Required to assist the evacuation crew in assessing the situation and determining the need for assistance. The evacuation vehicle often receives more definitive guidance while en route, such as the enemy's specific location to assist in planning an aircraft's approach.
	[in peacetime] Number and type of wounds, injuries, and illnesses	Specific information regarding any patient's wounds by type (for example, gunshot or shrapnel). Report serious bleeding along with the patient's blood type if known.	From evaluation of patient(s).	Medic or senior person present	Required to assist evacuation personnel in determining necessary treatments and special equipment.

Table B-1. Explanations of the nine lines on a medical evacuation request (continued)

Line	Item	Explanation	Where and How Obtained	Normal Provider	Reason
7	Method of marking pickup site	A—Panels B—Pyrotechnic signal C—Smoke signal D—None E—Other	From evaluation of situation and materiel availability.	Medic or senior person present	Required to assist the evacuation crew in identifying the specific pickup location. Do not transmit the panel or smoke color until the evacuation vehicle contacts the unit prior to arrival. For security, the crew identifies the color, and the unit verifies it.
8	Patient nationality and status	A—U.S. military B—U.S. citizen C—Non-U.S. military D—Non-U.S. citizen E—Enemy prisoner of war The transmission of the number of patients in each category is unnecessary.	From evacuation platform.	Medic or senior person present	Required to assist in planning for destination facilities and any need for guards. The unit requesting support ensures an English-speaking representative is present at the pickup site.
9	[in wartime] Chemical, biological, radiological, and nuclear contamination	C—Chemical B—Biological R—Radiological N—Nuclear Include this line only when applicable.	From evaluation of situation.	Medic or senior person present	Required to assist in mission planning. Determine which evacuation vehicle will accomplish the mission and when the mission will be accomplished.
	[in peacetime] Terrain description	Include details of terrain features in and around the proposed landing site. Describe the site relative to any prominent terrain features such as a lake, mountain, or tower.	From area survey.	Present personnel	Required for evacuation personnel to assess a route or avenue of approach into the area. Becomes particularly important in hoist operations.

Legend: #—number, SURG—surgery; U.S.—United States

Source Notes

This section lists sources by page number and in the order in which they appear in the publication.

xix, Ranger Creed: "Recognizing that I volunteered . . . .": Command Sergeant Major Neal Gentry, "Ranger Creed," https://www.benning.army.mil/Tenant/75thRanger/Ranger-Creed.html.

8-1, React to Direct Fire Contact While Dismounted – Platoon (07-PLT-D9501): "Conditions: The platoon is conducting . . . to maneuver the platoon.": Battle Drill (Drill Task) 07-PLT-D9501, 1–5.

8-4, React to Direct Fire Contact While Dismounted – Squad (07-SQD-D9501): "Conditions: The squad is conducting . . . to maneuver the squad.": Battle Drill (Drill Task) 07-SQD-D9501, 1–5.

8-7, Conduct a Platoon Assault (07-PLT-D9514): "Conditions: The platoon is conducting . . . to the company commander.": Battle Drill (Drill Task) 07-PLT-D9514, 1–6.

8-10, Conduct a Squad Assault (07-SQD-D9515): "Conditions: The squad is conducting . . . to the platoon leader.": Battle Drill (Drill Task) 07-SQD-D9515, 1–5.

8-13, Break Contact – Platoon (07-PLT-D9505): "Conditions: The platoon is conducting . . . designated rally point.": Battle Drill (Drill Task) 07-PLT-D9505, 1–5.

8-17, Break Contact – Squad (07-SQD-D9505): "Conditions: The squad is conducting . . . designated rally point.": Battle Drill (Drill Task) 07-SQD-D9505, 1–5.

8-21, React to Ambush (Dismounted) – Platoon (07-PLT-D9502): "Conditions: The platoon is conducting . . . reports the contact.": Battle Drill (Drill Task) 07-PLT-D9502, 1–5.

8-25, React to Ambush (Dismounted) – Squad (07-SQD-D9502): "Conditions: The squad is conducting . . . reports the contact.": Battle Drill (Drill Task) 07-SQD-D9502, 1–5.

8-29, Enter and Clear a Room – Squad (07-SQD-D9509): "Conditions: The squad is conducting . . . reorganizes, as needed.": Battle Drill (Drill Task) 07-SQD-D9509, 1–5

8-33, React to Indirect Fire While Dismounted – Platoon (07-PLT-D9504): "Conditions: The platoon is conducting . . . reports the contact.": Battle Drill (Drill Task) 07-PLT-D9504, 1–3.

8-34, React to Indirect Fire While Dismounted – Squad (07-SQD-D9504): "Conditions: The squad is conducting . . . reports the action.": Battle Drill (Drill Task) 07-SQD-D9504, 1–3.

This page intentionally left blank.

Glossary

The glossary lists acronyms with Army or joint definitions.

ACRONYMS AND ABBREVIATIONS

#	number
5-S	search, silence, segregate, safeguard, speed to rear
AA	avenue of approach
ADP	Army doctrine publication
AO	area of operations
AR	Army regulation
AT	antitank
ATP	Army techniques publication
BTC	bridge team commander
CAS	close air support
CASEVAC	casualty evacuation
CBRN	chemical, biological, radiological, and nuclear
CCP	casualty collection point
CLS	combat lifesaver
COA	course of action
DA	Department of the Army
DA Pam	Department of the Army pamphlet
DD	Department of Defense (form)
EPW	enemy prisoner of war
FDC	fire direction center
FM	field manual, frequency modulation
FO	forward observer
FRAGORD	fragmentary order
GEAR	gauge, evaluate, analyze, reduce
GOTWA	going, others, time, what, action
HE	high explosive
HLZ	helicopter landing zone
HQ	headquarters

IED	improvised explosive device
KIA	killed in action
LACE	liquids, ammunition, casualties, and equipment
LOA	limit of advance
LZ	landing zone
MARCH PAWS	massive bleeding, airway, respiration, circulation, hypothermia and head injuries pain, antibiotics, wounds, splints
MCO	Marine Corps order
MCRP	Marine Corps reference publication
MDI	modernized demolition initiator
MEDEVAC	medical evacuation
METT-TC (I)	mission, enemy, terrain and weather, troops and support available, time available, civil considerations, and informational considerations
MIST	mechanism of injury, injury type, signs, treatment
mm	millimeter
MOPP	mission-oriented protective posture
mph	miles per hour
NATO	North Atlantic Treaty Organization
NLT	not later than
NVD	night-vision device
OP	observation post
OPORD	operation order
PACE	primary, alternate, contingency, emergency
PB	patrol base
PIR	priority intelligence requirement
PL	platoon leader
PLT-D	platoon drill
PSG	platoon sergeant
PZ	pickup zone
R&S	reconnaissance and surveillance
RED	risk estimate distance
RP	release point
RTO	radio-telephone operator

SALUTE	size, activity, location, unit, time, and equipment
SE	site exploitation
SL	squad leader
SOP	standard operating procedure
SQD-D	squad drill
STANAG	standardization agreement
STP	Soldier training publication
SW	surface warfare
TC	training circular
TFC	tactical field care
TL	team leader
TLP	troop leading procedures
TM	technical manual
TO	technical order
TQ	tactical questioning
U.S.	United States
WARNORD	warning order
WP	white phosphorus
WSL	weapons squad leader

This page intentionally left blank.

References

All websites accessed on 15 July 2025.

REQUIRED PUBLICATIONS

These documents must be available to intended users of this publication.

DOD Dictionary of Military and Associated Terms. May 2025.

FM 1-02.1. *Operational Terms.* 28 February 2024.

FM 1-02.2. *Military Symbols.* 23 January 2025.

RELATED PUBLICATIONS

These documents are cited in this publication.

ARMY PUBLICATIONS

Most Army doctrinal publications are available online: https://armypubs.army.mil. Most battle drills are available online: https://rdl.train.army.mil/.

ADP 3-0. *Operations.* 21 March 2025.

AR 385-63/MCO 3570.1D. *Range Safety.* 23 May 2025.

ATP 3-06.11. *Brigade Combat Team Urban Operations.* 27 September 2024.

ATP 3-09.32/MCRP 3-31.6/NTTP 3-09.2/AFTTP 3-2.6. *Multi-service Tactics, Techniques, and Procedures for Joint Application of Firepower.* 29 November 2023.

ATP 3-21.8. *Infantry Rifle Platoon and Squad.* 11 January 2024.

ATP 4-01.45/MCRP 3-40F.7/NTTP 4-01.6/AFTTP 3-2.58. *Multi-service Tactics, Techniques, and Procedures for Tactical Convoy Operations.* 26 March 2021.

ATP 4-02.2. *Medical Evacuation.* 12 July 2019.

ATP 5-19. *Risk Management.* 9 November 2021.

Battle Drill (Drill Task) 07-PLT-D9501. *React to Direct Fire Contact While Dismounted–Platoon.* 26 July 2022.

Battle Drill (Drill Task) 07-PLT-D9502. *React to Ambush (Dismounted)–Platoon.* 26 July 2022.

Battle Drill (Drill Task) 07-PLT-D9504. *React to Indirect Fire While Dismounted–Platoon.* 26 July 2022.

Battle Drill (Drill Task) 07-PLT-D9505. *Break Contact–Platoon.* 26 July 2022.

Battle Drill (Drill Task) 07-PLT-D9514. *Conduct a Platoon Assault.* 30 August 2022.

Battle Drill (Drill Task) 07-SQD-D9501. *React to Direct Fire Contact While Dismounted–Squad.* 26 July 2022.

Battle Drill (Drill Task) 07-SQD-D9502. *React to Ambush (Dismounted) Squad*. 26 July 2022.

Battle Drill (Drill Task) 07-SQD-D9504. *React to Indirect Fire While Dismounted Squad*. 26 July 2022.

Battle Drill (Drill Task) 07-SQD-D9505. *Break Contact Squad*. 26 July 2022.

Battle Drill (Drill Task) 07-SQD-D9509. *Enter and Clear a Room Squad*. 11 July 2022.

Battle Drill (Drill Task) 07-SQD-D9515. *Conduct a Squad Assault*. 11 July 2022.

FM 6-27/MCTP 11-10C. *The Commander's Handbook on the Law of Land Warfare*. 7 August 2019.

FM 7-0. *Training*. 14 June 2021.

TC 3-22.19. *Grenade Machine Gun MK19 Mod 3*. 10 May 2017.

TC 3-22.50. *Heavy Machine Gun M2 Series*. 19 May 2017.

TC 3-22.240. *Medium Machine Gun*. 28 April 2017.

TC 3-22.249. *Light Machine Gun M249 Series*. 16 May 2017.

TC 3-97.61. *Military Mountaineering*. 26 July 2012.

TM 3-34.82/MCRP 3-34.2. *Explosives and Demolitions*. 7 March 2016.

TM 9-1005-201-10/Air Force TO 11W3-5-5-51/Marine Corps TM 08671A-10/1B. *Operator Manual for Machine Gun, 5.56mm, M249 w/Equip; (NSN 1005-01-127-7510) (EIC: 4BG) (AR Role); (NSN 1005-01-451-6769) (EIC: 4BK) (LMG Role)*. 31 May 2019.

TM 9-1005-213-10/Air Force TO 11W2-6-3-161/Marine Corps TM 1005-10/1/NAVSEA SW360-AW-OPI-010. *Technical Manual for Machine Gun, Caliber .50: M2A1 with Fixed Headspace and Timing (NSN 1005-01-511-1250) (EIC: 4AZ)*. 21 August 2017.

(CUI) TM 9-1005-313-10/Air Force TO 11W2-6-5-1/Marine Corps TM 09712B-10/Navy SW360-AH-OPI-010 Rev 1. *Operator Manual for Machine Gun, 7.62 mm, M240 (NSN 1005-01-025-8095) P/N 11826290 (EIC 4BD)*. 15 May 2022.

TM 9-1010-230-10/Air Force TO 11W2-5-16-1/Marine Corps TM 08521A-OR/1/Navy SW363-C3-MMM-010. *Technical Manual Operator's Manual for Machine Gun, 40 mm, MK19 Mod 3 (NSN 1010-01-126-9063) (EIC 4AE)*. 31 August 2012.

(CUI) TM 9-1010-233-10. *Operator Manual for Lightweight Company Mortar, 60mm, M224A1 (NSN 1010-01-586-2874) (EIC 4ST)*. 30 July 2024.

TM 9-1015-249-10/Marine Corps TM 09922A-10/1. *Technical Manual Operator's Manual for Mortar, 81 mm, M252 (NSN 1015-01-164-6651) (EIC 4SK)*. 10 August 2012.

TM 9-1015-257-10. *Technical Manual Operator Manual for Mortar, 81-mm, M252A1 (NSN 1015-01-586-2135) (EIC: 4SU)*. 25 August 2014.

OTHER PUBLICATIONS

Standardization Agreements are available online: https://nso.nato.int/nso/nsdd/main/list-promulg.

STANAG 2546. *Allied Joint Medical Doctrine for Medical Evacuation*, Edition 2. 29 August 2018.

STANAG 3204. *Aeromedical Evacuation*. 6 July 2020.

WEBSITES

Army Training Network. https://atn.army.mil/.

Department of Defense Law of War Manua. https://www.defense.gov/.

First Geneva Convention: Convention (I) for the Amelioration of the Condition of the Wounded and Sick in Armed Forces in the Field, 12 August 1949. https://www.icrc.org/en/war-and-law/treaties-customary-law/geneva-conventions.

PRESCRIBED FORMS

This section contains no entries.

REFERENCED FORMS

Unless otherwise indicated, DA forms are available online: https://armypubs.army.mil. DD forms are available online: https://www.esd.whs.mil/Directives/forms.

DA Form 1156. *Casualty Feeder Card.*

DA Form 2028. *Recommended Changes to Publications and Blank Forms.*

DA Form 5517. *Standard Range Card.*

DD Form 1380. *Tactical Combat Casualty Care (TCCC) Card.* Printed forms are available through normal forms supply channels.

RECOMMENDED READINGS

These documents contain relevant supplemental information not cited in the publication. Most Army doctrinal publications are available online: https://armypubs.army.mil.

ADP 6-22. *Army Leadership and the Profession.* 31 July 2019.

AR 350-1. *Army Training and Leader Development.* 1 June 2025.

ATP 3-09.42. *Fire Support for the Brigade Combat Team.* 1 March 2016.

ATP 3-18.12. *Special Forces Waterborne Operations.* 16 December 2022.

ATP 3-18.14. *Special Forces Ground Mobility Operations Tactics, Techniques, and Procedures.* 15 July 2022.

ATP 3-21.50. *Infantry Small-Unit Mountain and Cold Weather Operations.* 27 August 2020.

ATP 3-75. *Ranger Operations.* 26 June 2015.

ATP 6-02.53. *Techniques for Tactical Radio Operations*. 13 February 2020.

ATP 6-02.72/MCRP 3-30B.3/NTTP 6-02.2/AFTTP 3-2.18. *TAC Radios: Multi-Service Tactics, Techniques, and Procedures for Tactical Radios*. 14 July 2021.

FM 3-09. *Fire Support and Field Artillery Operations*. 12 August 2024.

FM 6-02. *Signal Support to Operations*. 13 September 2019.

FM 6-99. *U.S. Army Report and Message Formats*. 17 May 2021.

STP 21-1-SMCT. *Soldier's Manual of Common Tasks Warrior Skills Level 1*. 16 October 2023.

TC 3-25.26. *Map Reading and Land Navigation*. 15 November 2013.

TC 8-800. *Medical Education and Demonstration of Individual Competence*. 15 December 2021.

TC 18-35. *Special Forces Tracking and Countertracking*. 11 May 2023.

TM 3-22.31. *40-mm Grenade Launchers*. 17 November 2010.

Journals of Major Robert Rogers: Reprinted from the original edition of 1765. Henry Bamford Parkes, New York: Corinth Books. 1961.

SOURCES USED

Ranger Creed, Command Sergeant Major Neal Gentry, 1974. https://www.benning.army.mil /Tenant/75thRanger/index.html.

Index

Entries are by paragraph number.

TC 3-21.76

19 September 2025

TC 3-21.76
19 September 2025

By Order of the Secretary of the Army:

RANDY A. GEORGE
General, United States Army
Chief of Staff

Official:

MATTHEW L. SANNITO
Administrative Assistant
to the Secretary of the Army
2525844

DISTRIBUTION:
Active Army, Army National Guard, and United States Army Reserve. To be distributed in accordance with the initial distribution number (IDN) 116076, requirements for TC 3-21.76.

This page intentionally left blank.

This page intentionally left blank.

This page intentionally left blank.